Taking

Nature

Into

Account

Taking

A Report

Nature

to the

Into

Club of Rome

Account

Toward a Sustainable National Income

WOUTER VAN DIEREN, EDITOR

COPERNICUS
AN IMPRINT OF SPRINGER-VERLAG

© 1995 Springer-Verlag New York, Inc.

Published in the United States by Copernicus, an imprint of Springer-Verlag New York, New York, Inc.

Copernicus
Springer-Verlag New York, Inc.
175 Fifth Avenue
New York, NY 10010

Library of Congress Cataloging in Publication Data

Taking nature into account: a report to the Club of Rome / Wouter van
Dieren, editor.
    p.   cm.
    Includes bibliographical references (p.   ) and index.
    ISBN 0-387-94533-4 (softcover)
    1. National income—Environmental aspects—Accounting.
    2. Sustainable development—Accounting.   3. Externalities
    (Economics)—Accounting.   4. Economic indicators.   I. Dieren,
    Wouter van.   II. Club of Rome.
    HB141.5.T35   1995
    333.7—dc20   95-15974

Manufactured in the United States of America.
Printed on acid-free paper.

9 8 7 6 5 4 3 2 1

ISBN 0-387-94533-4

Contents

List of Contributors

- Hans Achterhuis, University of Twente, the Netherlands
- Bart de Boer, Central Bureau of Statistics, the Netherlands
- Peter Bosch, Central Bureau of Statistics, the Netherlands
- Herman Daly, Maryland University, former World Bank economist, United States
- Erik van Dam, Institute for Environment and Systems Analysis, the Netherlands
- Wouter van Dieren, Club of Rome, director of Institute for Environment and Systems Analysis, the Netherlands
- Paul Ekins, University of London, United Kingdom

- John Elkington, director of SustainAbility, United Kingdom
- Salah El Serafy, freelance consultant, former World Bank economist, United States
- Robert Goodland, World Bank, United States
- Hazel Henderson, publicist for U.S. Office of Technology Assessment, United States
- Roefie Hueting, Central Bureau of Statistics, the Netherlands
- Ronald Janssen, Central Bureau of Statistics, the Netherlands
- Alexander King, Club of Rome, France
- Christian Leipert, freelance researcher, former Wissenschaftszentrum für Sozial-Forschung, Germany
- Cees Oomens, Central Bureau of Statistics, the Netherlands
- Sandra Postel, director, Global Water Policy Project, United States
- Emil Salim, former Minister of State Environment and Population, Indonesia
- Eberhard Seifert, Wuppertal Institut für Klima, Umwelt und Energie, Germany
- Ismail Serageldin, vice-president World Bank, United States
- Fulai Sheng, economist, World Wide Fund for Nature, Switzerland
- Jan Paul van Soest, director Centre for Energy Conservation and Environmental Technology, the Netherlands
- Carsten Stahmer, Statistisches Bundesamt Wiesbaden, Germany
- Ernst von Weizsäcker, Club of Rome, director of Wuppertal Institut für Klima, Umwelt und Energie, Germany

Foreword

This report is about a major necessary correction factor in the world economy: the need to calculate environmental degradation, in the widest sense, within the system of national accounts.

Every since the Club of Rome published *Limits to Growth,* we have pleaded for information systems on the world economy that produce correct data, proper figures, and honest predictions. Growth, as we have defined it, is the expansionist activity of ever more production, activities which are formally measured as the gross national product (GNP), the outcome of the accounts. However, economic growth in the formal definition of welfare theory is the reduction of scarcity.

Hence, if production is creating scarcity rather than reducing it, economic growth is negative.

Over time, many economists have pointed at the risk for the future of the world if indeed the use of natural resources is not accounted for as a loss, but on the contrary, as income. Yet, this is today's practice: consumption, abuse, and even depletion of natural resources—the very basis of life—is formally registered as growth and income.

To a large extent, the early work of Dr. Roefie Hueting—whose *New Scarcity and Economic Growth* (1974) is a milestone in pointing at the paradoxes of economic theory and daily use—paved the way for a mainstream of economists who now advocate the need for drastic change. In this report, we often refer to Hueting, who is now studying the corrected national income, when we discuss research into the conditions of a sustainable demand-and-supply systems theory.

The editor of this report, Wouter van Dieren, chairman of the Institute for Environment and Systems Analysis, has been largely responsible for the success of the publicity around *Limits to Growth*. Ever since, he has worked as an environmental-economics pioneer, as such publishing *Nature's Price* in 1977 (together with M.G.W. Hummelinck), based partly on the work of Hueting, a book for the World Wildlife Fund (now the World Wide Fund for Nature).

He has invited a range of the best environmental economists around the world to contribute to this report. In the Preface, the specific contributions of each specialist are described.

It is our wish that this strong plea for changing the prime information system of the economy is heard globally.

We have to know the facts. We cannot create a sustainable future it we keep dragging a veil over reality, not only ignoring depletions and the collapse of life support systems, but actually counting this as progress. The limits to growth will then hit us even faster.

Dr. Alexander King

Preface

STRUCTURE OF THE REPORT

This report concerns a debate that threatens to be dominated by economists although it defines the reality of us all. The structure of the report therefore aims to guide the noneconomist through this debate and to provide him or her with the intellectual ammunition to enter into it.

The Introduction describes our motivation for publishing this report. The limits to growth have been reached without us even noticing it. Society relies on economic indicators (GDP, national income) that provide distorted information (Chapter 1).

Part I begins by unravelling the philosophical and societal roots that may explain humanity's fatal attraction to material progress and economic growth (Chapter 2). Next, the establishment of the System of National Accounts, the system used to record economic growth, is reconstructed. We present its architects and its original assumptions and aims. Then we document the discrepancies between these assumptions and goals and contemporary reality (Chapter 3).

Part II deals with the paradoxes of economic growth, mirrored in the state of the environment and the quality of life in many parts of the world, and the flaws in the indicators that should be informing us of this. We take a broad view of the alarming state of the earth's carrying capacity (Chapter 4). After pointing out the consequences of giving the highest priority in economic policy to eternal growth of production, the failure of the System of National Accounts to give us the right signals on both the economy (production) and welfare is explained in detail (Chapter 5).

Sustainable development, society's answer to these paradoxes, is examined in Part III. The history of the process of sustainable development and an overview of the roles of the actors in the process (e.g., multilateral organizations and NGOs) are given (Chapter 6). The practical implications of the concept of sustainability for environmental, economic, and social policies are discussed (Chapter 7). Special attention is given to the meaning of sustainable development for the countries of the southern hemisphere (Chapter 8). Part III closes with a description of the role of indicators in monitoring sustainable development and a view of the range of different "beyond GDP" approaches—their purpose, level of aggregation, valuation methodology, possibilities, and limitations (Chapters 9 and 10).

In Part IV, leading experts in the field of environmental accounting present their views on the conceptual and methodological aspects of adjusting GDP for changes in the value of natural capital due to economic activity (Chapters 11, 12, 13). In Chapter 14 we refer to the work of Carsten Stahmer, the main author of the *U.N. Handbook on Integrated Economic and Environmental Accounting*, assessing and putting into context the approaches of the other experts. The differences and similarities of the various proposals are made explicit, and their strengths and weaknesses are discussed.

In Part V, we make specific recommendations concerning:

- the improvement of welfare measurement;
- the improvement of GDP as a measure of economic output;

- the anticipated resistance to the innovations proposed in the report by statistical offices and politicians; and
- the way to strengthen international cooperation among experts.

The recommendations in this book are consistent with the ones in "Real Value for Nature" by WWF International (Sheng, 1995). *Taking Nature Into Account* was supported by the Dutch ministries for the environment and for foreign affairs, and by the World Wide Fund for Nature.

ACKNOWLEDGMENTS

This report is the outcome of a venture in which many individuals and institutions from all over the world have taken part. Each of them has provided parts of the story this report tells. My role has been to make sure that these parts blend into one coherent vision.

Hans Achterhuis is Professor of Eco-philosophy at the University of Twente, the Netherlands. In his book *Het Rijk van de Schaarste* (The Reign of Scarcity), Achterhuis exposed in unparalleled fashion the concept of scarcity as the founding myth of modern social thought. It was for this reason that I asked him to provide input for Chapter 2, "The History and Roots of Growth."

Erik van Dam is research fellow at the Institute for Environment and Systems Analysis in Amsterdam, member of the editorial team of this book, and principal author of Sections 1–7 of Part V.

Wouter van Dieren is editor of this book, director of the Institute for Environment and Systems Analysis in Amsterdam, member of the Club of Rome, vice-president of the Advisory Board of the Wuppertal Institute for Climate, Environment, and Energy, and member of the World Academy for Art and Science. He stood at the cradle of the Dutch Environmental Movement and, as publisher and television producer, he has written and produced hundreds of articles and programs on the themes of nature, the environment, and science.

Paul Ekins has inspired a number of initiatives that aim at a fundamental reform of economics, and has edited or written several green economy–related books such as *Real Life Economics: Understanding Wealth Creation* (1992) and *Wealth Beyond Measure: An Atlas of New Economics* (1992). He is currently Research Fellow at the Department of Economics, Birkbeck College, University of London. Ekins uncovers the failures of the economic information system of our time, which he does in Chapter 5, "Failures of the System of National Accounts."

I asked **John Elkington,** one of the most creative and productive missionaries of sustainability in practice, to focus his efforts on describing the sustainable development process in Chapter 6. I am grateful to him for conscientiously composing this necessary but difficult building block of the report.

Hazel Henderson is an internationally established futurist, lecturer, and consultant on alternative developments, advising corporations, governments, and NGOs in over 30 countries. A member of the Board of Directors of the Worldwatch Institute, and Founder of the Center for Sustainable Development and Alternative World Futures, her previous works include *Creating Alternative Futures: The End of Economics* (1978), *Politics of the Solar Age: Alternatives to Economics* (1981), and *Paradigms in Progress: Life Beyond Economics* (1993). Hazel Henderson is one of the indisputable pioneers of alternative economics. Her criticism of the dominance of economic thought is provocative and forceful, and she helped a great deal with Chapter 9.

Roefie Hueting of the Netherlands Central Bureau of Statistics is to a great extent not only responsible for Chapter 13, but also for the original baseline of this report, as Alexander King explains in the Introduction. In 1974, Hueting published *New Scarcity and Economic Growth,* marking the start of lengthy discussions on the need to adjust the Gross National Product for environmental losses. Throughout the last 20 years, Hueting has been a key researcher in this field. His present work is concerned with solving the methodological problems relating to determination of the National Income in a (hypothetical) environmentally sustainable economy. **Bart de Boer** and **Peter Bosch** are researchers at the Netherlands Central Bureau of Statistics, and as such are responsible for substantiating the original Hueting concept, a job which they have done with enthusiasm for several years. Together with Hueting and **Jan Paul van Soest** they worked on Chapter 13. Hueting bears the final responsibility for the text, which has been edited by Jan-Paul van Soest. Van Soest is director of the Center for Energy Conservation and Environmental Technology in Delft and one of the key actors in the energy/economics debate in the Netherlands.

Christian Leipert was one of the first researchers to undertake conceptual and empirical work on defensive expenditures. At the moment he is working with the Ecological Economic Research Institute in Berlin. He has also been engaged with the Wissenschaftszentrum für Sozial-Forschung in Berlin. In Chapter 11, he expresses his vision of defensive expenditures in relation to GDP/NDP.

Cees Oomens, together with Jan Tinbergen, has been closely involved with the creation and development of both the Dutch and the International System of National Accounts (SNA). His dedication to his work is exemplified by the fact that, despite his long retirement and great age, until recently he was most readily reached at the office that the Central Bureau had kept for him. **Ronald Janssen** is head of the Division for Quarterly Accounts (Department for National Accounts) of the Netherlands Central Bureau of Statistics. He is author of several textbooks on the SNA and is known for his gift for making this subject accessible to nonstatisticians. The combined expertise of Cees Oomens and Ronald Janssen provides a chapter that unlocks the SNA to the public and provides insight into the period of creation of the system (Chapter 3).

Sandra Postel is Director of the Global Water Policy Project in Cambridge, Massachusetts, and Adjunct Professor of International Environmental Policy at Tufts University. For six years, she served as vice-president for research at the Worldwatch Institute, with which she remains affiliated as senior fellow. She is co-author of 11 of Worldwatch's "State of the World" reports, and author of *Last Oasis: Facing Water Scarcity.* My charge to her was to irrefutably describe the state of the environment, without flooding the reader with endless amounts of figure-loaded tables for proof. She has succeeded extremely well in Chapter 4.

Emil Salim, former Minister of Environment and Population, former member of the World Commission on Environment and Development, and now a member of the Commission on Sustainable Development, assisted in writing Chapter 9. For this chapter, I also drew heavily on the extensive research of **Martin Khor Kok Peng,** who is a renowned publicist on North-South relations and Director of the Third World Network in Penang, Malaysia.

On behalf of the Wuppertal Institut für Klima, Unwelt and Energie, **Eberhard Seifert** is working on new models of wealth. His contribution to this book is to be found in "The Role of Social Indicators" (Chapter 9), in which we give an overview of developments in the field of welfare indicators. I have much appreciated the fruitful discussions he and I have had on this theme.

As a World Bank economist, **Salah El Serafy** has been involved in the debate on sustainable national income for many years. As editor of, and contributor to, the milestone World Bank/UNEP publications on environmental accounting, and as a keynote speaker at many conferences, he has been a very strong and inspiring advocate of adjusting GDP for the consumption of natural capital. Besides stimulating and contributing to the debate on proper national income accounting in general, he has developed and put forward his own approach to dealing with the depletion of natural resources: the user cost approach. Both of these topics are dealth with in very clear language in Chapter 12.

As vice-president for environmentally sustainable development, **Ismail Serageldin** has been involved in the latest World Bank publication, *Towards Improved Accounting for the Environment*. Both **Robert Goodland** and **Herman Daly** have been pioneers in developing and stressing the ideas of sustainability within the World Bank. We are very glad to have the three of them contribute to this report and especially to Chapter 7. Herman Daly, author of *Steady-State Economics* and *For the Common Good*, is now teaching at Maryland University.

Fulai Sheng is economist for the Sustainable Resource Programme of the World Wide Fund for Nature International (WWF-I) in Gland, Switzerland. He became involved in the subject when, in 1992, WWF-I published Roefie Hueting's *Methodology for the Calculation of Sustainable National Income*. By the end of 1993, I got to know Fulai and his work. In that period, WWF-International was starting further activities in the field of environmental accounting and the SNA reform. This contact led to close cooperation between IMSA and WWF-International. Today, Fulai attends most meetings and conferences on national accounting and indicators for sustainable development, representing WWF's view on indicators, environmental accounting in general, and the *U.N. Handbook on Integrated Economic and Environmental Accounting* in particular. WWF-International has just published "Real Value for Nature" (Sheng, 1995), a survey of environmental accounting which contains a comprehensive overview of environmental accounting that renders this topic much more accessible to both experts and nonexperts alike. Fulai has been a very important adviser, particularly for Parts IV and V of this report, to which he contributed with a plea for strengthening international coordination.

Carsten Stahmer, from the Statistisches Bundesamt in Wiesbaden, Germany, has been highly involved in developing the United Nations System for Integrated Environmental and Economic Accounting (SEEA), which aims at providing a consistent framework for national statistical offices to set up their environmental accounting systems. Carsten has spent two years in the United States working as a consultant for UNSTAT. Together with Peter Bartelmus and Jan van Tongeren he has written the *U.N. Handbook on Integrated Economic and Environmental Accounting,* which was published in 1993. In his opinion, the main task of the *Handbook* is to effect a synthesis of the approaches of the different schools of thought in the field of natural resources and environmental accounting. Carsten was one of the main advisers for Part V of this report and has provided a summary of the *U.N. Handbook* and an assessment of valuation methodologies in Part V. I am very grateful for having had the opportunity to collaborate with him, both for his excellent knowledge and expertise, and for his patience and friendship.

Ernst Ulrich von Weizsäcker is president of the Wuppertal Institut for Climate, Environment and Energy, and member of the Club of Rome. He is co-responsible for Section 8 of Part V, in which both the ideas of the Club of Rome on an adjusted national income are presented, and the direction innovations have to take is described. During these last years, Ernst's principal work has been to reorient our economic system and the goals of our philosophy of life, and he has explicitly asked for new welfare models. His work has been an important impulse to publish this book.

I have also made extensive use of the works of Club of Rome members **Orio Giarini** and **Manfred Max-Neef,** whose pioneering efforts in this field gave momentum to global thinking on wealth, welfare, and GDP.

This book is the result of a process in which many of the contributors took part intensively together with myself and my colleagues at IMSA. In this process, many of us have brainstormed together, visited one another, telecommunicated and quarrelled, learned from one another, and shared uncertainties and convictions. As a result, this book reflects the strong views held by many of the authors. There is a near-total consensus on every conclusion.[1] Along the way an alliance was

---

[1] Not all of the authors necessarily agree with every conclusion.

forged, because it turned out that many of us have many things in common:

- audacity, because of the tremendous forces of vested interests; every day we experience those forces and experience tension between the envisioned world and the real one;
- imagination, because we seemingly dare to envision as a given the modification of a system that is considered to be a good reflection of economic reality, and thus as something that is not to be questioned;
- the creativity to think of strategic ways to deliver the message;
- a sense of humor;
- the quest for a balance between scientific depth and accuracy and the necessity to end dialectics in order to come to practical solutions;
- integrity;
- involvement;
- a sense of urgency: no time to waste, many of the authors strongly feel that the carrying capacity of the earth is seriously challenged;
- optimism; the time is ripe; this is not the beginning; it is part of the enactment of the transformation of society.

The IMSA staff consisted of the following: Heske van Veen coordinated the editing process of Parts I, II, and III. Erik van Dam was responsible for the coordination of the editing process of Part IV and the writing of Sections 1–7 of Part V, part of which was executed in cooperation with Hans Taselaar from GAIA Ecological Economics. David Rosenberg made a solid and fast contribution to Chapter 10. Reynt-Jan Sloet van Oldruitenborgh managed the team, which also consisted of Marion von Richter (external reltions, fund raising), Eva Klok, who did the majority of the typing and the layout of the different versions of the manuscript, Kathelijne Drenth (external relations), and Nigel Harle (English editor). All of them have worked for two years on this extensive, demanding task, for which I am deeply grateful to them. Without Heske van Veen and Erik van Dam, particularly, the job could not have been done.

Wouter van Dieren

CHAPTER 1

Introduction

In 1972, a report was published by a mysterious club nobody had ever heard of, which shocked the world—a report about the forthcoming collapse of life on earth, not written by some sectarian doomsday prophet but by scientists of high repute, working with that new device of modernity, the computer.

"Limits to Growth," the report was called, and besides a shock it also caused outrage worldwide. Several years after the first phase of environmental awareness and shortly before the first oil crisis (1973), "Limits" brought the message that the world was heading for disaster because of unfettered population growth and industrial expansion, exhaustion of stocks of natural resources, environmental destruction, and food shortages.

"Limits" was based on a so-called simulation model, a mathematical representation of the main variables and their dynamic interactions known as the WORLD III model. Some of the key features of these dynamics are feedback loops, which show that an intervention in one part of a system has unexpected effects on other variables of that system.

The forms of exhaustion predicted in the various scenarios simulated in the model start to emerge in the early twenty-first century, as the world population grows to a peak of 10 billion, per capita food production drops to a mere 15–25 percent of 1970 levels, pollution has risen tenfold, and the most important resources, such as oil and gas, have become depleted. Because of the so-called exponential character of growth and depletion, half-hearted or one-sided measures are of little avail. A drastic program of technological improvement such as energy conservation, for example, achieving 50 percent savings in 20 years against a background of, say, 2 percent growth in consumption, postpones the date of depletion by a mere 3 years.

"Limits" became the subject of heated controversy, and the Club of Rome soon gained the reputation of being a neo-Malthusian movement of doomsayers. The report became world famous, an indication that its message was not only controversial but also supremely recognizable.

Although thousands of scientists have devoted their efforts to the question of how reliable WORLD III was and whether it is even at all possible to forecast the future in this manner, "Limits" has, in our view, come through all the criticism untarnished. In the first place, because the primary aim was not to make a prediction but "to improve the insight," in the words of Jay Forrester, one of the contributing authors; and secondly, because nobody has yet really succeeded in finding fault with the main calculations and the underlying hypotheses.

Since 1972, countless studies and books have been published that confirm the message of "Limits"; but even more extensive than this scientific work has been the worldwide denial of the limits to growth, and the impassioned attempts to remain one step ahead of the imminent shortages through policies of continued economic expansion. Meanwhile, additional new insights have arisen, which not only confirm the impending disasters but also indicate that the limits to growth may well have been exceeded and that the world has been in a state of decline for some years already. The most important study in this context is For the Common Good (1989), in which Daly and Cobb develop an information theory to replace or supplement the incomplete data function of what is known as the Gross National Product. By processing U.S. statistical data on some twelve so-called welfare indicators, they drew the conclusion that for the last twenty years the link between production growth and the creation of welfare has become progressively weaker; prior to that date, production growth had achieved exactly what Adam Smith foresaw in 1752: the addition of value so as to indeed create the "Wealth of Nations." In the 1970s this link began to be lost, however, and this process is proceeding at such an accelerating pace that we are now confronted with the curious phenomenon of production growth leading to a decline in welfare; stated differently, the limits to growth have been reached without us even noticing it, because we have been interpreting the figures wrongly.[1]

Our aim is to undertake a further analysis of the concept of growth within this framework; and above all, of the opposition to information that is critical of this growth. "Limits" evoked extreme opposition, but nonetheless there is fascination worldwide with the Club of Rome and its first and greatest message. The latest insights tell us that their message is not only being confirmed daily, but that it has in fact been surpassed by reality. The limits to growth no longer merely lie ahead, in the future; they are with us today, and have been for the last twenty years.

The international community does not know how to handle this reality, however. Discussions in international forums where the necessity of growth or its significance are challenged inevitably lead to emotional scenes, and all the agendas of all the world's political and economic bodies call for growth in the restricted sense of the word, because of an unabashed conviction that this is always good for the world. But this is simply not the case: a steady growth of output does not necessarily lead to more jobs or a better environment, it does not combat famine or promote social security, neither does it improve

---

[1] It is not our intention here to discuss the work of Daly and Cobb in any depth. The incompleteness of their methodology is also not the subject of this chapter. Their method has been tried in a number of countries, with similar results; the conclusions drawn for the U.S. are found to hold for the U.K., Germany, the Netherlands, and Austria, too. For details, we refer the reader to the literature (see the References).

education or public health. On the contrary, most of these aspects of welfare seem to suffer under unrestricted economic expansion, which has become a law unto itself.

The main thrust of opposition to "Limits" lies in the belief that economic growth is a kind of law of nature, which humanity must obey. Since Adam Smith invented the Invisible Hand, this power has been a guiding principle for all those who believe that free trade, or the market, will ultimately lead to a natural order of things, a moment when everything will fall neatly into place: free trade will provide income and employment, welfare for all, equality, peace, and a future. In this way of thinking, the problems outlined by "Limits" are a result of obstacles to free trade—and if things are not well with the world, that is a logical consequence of these obstacles; for example, too much government intervention, too high social benefits, too much environmental and labor legislation, an overly expensive quaternary sector, and so on. Allow the free market to do its curative work, in other words, and the Divine ordination of the invisible hand will balance out the world economy.

It is no coincidence that this kind of metaphysical notion was a nursemaid to the industrial revolution, nor that it is part and parcel of modern economics. Adam Smith certainly intended the Invisible Hand to serve as a metaphysical, divine principle, which effortlessly took over the role of Divine Providence, on which western humanity had focused its aspirations until the Enlightenment. The Enlightenment blocked this Providence, because it called for science, technology, and mechanization, and thus distracted attention from God's will. By introducing the invisible hand, Smith took up the deistic thread once more; now the economy too, or precisely the economy, was to be driven by supernatural laws, and in the industrial age, too, the role of God would remain of decisive importance.

It is our conviction that this metaphysic is still as topical as ever. The opposition to "Limits" is so strenuous that clearly forces other than science are at work. One would expect humanity to take up the challenges of "Limits" and set up an international organization to halt the decline. The opposite has been the case. A veritable crusade of economic expansionism has been unleashed, as if to prove that "Limits" was pessimistic and in error, and everywhere the conquests of this crusade are praised as providing the desired proof, such as the economic miracles embodied in the Asian growth figures. And for the sake of convenience we then ignore the enormous price of these miracles, the ecological destruction, the plundering of the surrounding oceans, the consumption of the region's natural capital, the underpaid workers, the absence of social security. And while these miracles are seen as proof of the power of the invisible hand, nobody is prepared to answer the question of why the same metaphysic has caused war and famine in Africa. Does the invisible hand pick its favorites? Or are the Africans paying the penalty of disobeying the laws of natural economic ordination? Or is it the case that here—and in the former Soviet Union—the law of Keynes holds: that suffering is a precondition, albeit temporary, for later success?

It is of crucial importance to state that the invisible hand does not exist, that there are no laws of economic ordination, that although the notion of economic growth can be defined, its political usage is above all rhetorical, that economics is not really a science but a set of theories, and that every attitude towards the limits to growth is a question of culture, of choices, free will, and—possibly—rationality. There is no inevitable fate compelling humanity towards unlimited free trade, over-exploitation of nature and labor, exhaustion of resources, and finally towards a war of all against all (Hobbes) to gain control of the last remaining resources and food. Economic thought differs from culture to culture, and within each culture even from school to school, from university to university. There are myriad options to choose from, and none of them need satisfy a single requirement of a metaphysical nature. What is of key importance is that we rid our economies of hypocrisy, and this forms the subject of the present Report.

The main hypocrisy lies in the system of National Accounts that has been employed in the Western economy for nearly half a century now, with partial implementation in most other countries. The technical background of the National Accounts is described elsewhere in this Report; in this introduction, our aim is simply to set out why we consider this topic to be so vitally important in the debate on growth and the limits to which it is subject.

Until 1945, the notion of economic growth was used differently from today. As elaborated in Chapter 3, it was not until about 1932 that several economists came up with the idea of measuring a country's economic performance and not until 1950 that the ensuing system was introduced in most industrialized countries. It was thus inevitable that the costs of production growth would be encountered, costs that for decades the theory had termed negative external effects. In former times these effects had been happily accepted, but when production as a whole is encapsulated in a profit-and-loss account, the costs, or negative expenditure, automatically appear on the balance sheet. And that is where we stand today.

Surprisingly enough, users of National Accounts have long remained deaf to recommendations to subtract these costs from the profits, despite an information load that has become so heavy in recent years that for some economies the point now appears to have been reached whereby the costs are perhaps even greater than the profits—without this being reflected in the National Accounts. This is of vital importance for the debate on the limits to growth, because these economies continue to literally count themselves rich, while poverty is on the rise, or in other words because the subtracted value is higher than the added value. Phrased differently: the economy is being kept afloat on paradoxical information, not even on incompleteness, and the abuse of the National Accounts is at the core of the matter.

The notion of the National Account is not employed in everyday political parlance; instead, the public at large hears the terms Gross National Product (GNP) and its derivative, Gross Domestic Product (GDP). This GNP has gained metaphysical significance: it stands for the mark given to the country by the Invisible Hand and thus even acts as a symbol of the degree to which that nation has been elected in the Divine ordination that steers the Invisible Hand. In this vision of things, one has subjected oneself to the natural laws of the economy, and the nation is seen to have passed the examination if the number attached to the GNP is positive: one, two, or three percent growth per annum, whatever that may mean. Orio Giarini has made a comparison between the effect of GNP in heaven, in hell, and on earth. He describes the complication of Industrial Revolution accounting by the paradox of hell and heaven, when applied to the notion of scarcity. Heaven, being probably blessed by an infinite stock of goods and services of all sorts (material and spiritual), knows nothing of scarcity. Economics and the economy therefore do not exist. There are no prices and there is no money, since everything is readily available without any restriction or work. Heaven, then, must be something very different from earth, but it is also a place of zero GNP. Hell, as the opposite of heaven, is a place which consumes a lot of energy in maintaining its celebrated image and presumed activities. It therefore probably needs to develop a huge value added which nobody has ever tried to measure: GNP must be very high indeed! On our earth, the maximum possible achievement in the fight against scarcity is to create an abundance in as many sectors as possible. But human and economic development also entails identifying and coping with new scarcities. Scarcity is ultimately the hallmark of the system of disequilibrium within which human endeavour is destined to operate: it is the *sine qua non* of man's quest for fulfilment, so Giarini says.

One of the major paradoxes in value accounting and in defining the development of wealth is that an increase in real wealth corresponds in some cases merely to an increase in the cost of pollution control (e.g. investment for waste-disposal and environmental purposes, which is clearly a deducted value type of cost), while on the other hand, many real increases in value are underrated. For instance, GNP growth figures published each year by governments indicate that the economy has grown by so many percent. However, a large part of this growth is in fact absorbed by factors which do not necessarily add to our wealth, while other factors that represent net increases in our well-being are not, or are only inadequately, taken into account.

Going back to the paradox of hell and heaven, one of the reasons for our reluctance to reconquer paradise is that in some weird way we seem to be more at ease with hell.

Giarini believes it important to define a level for the wealth of nations in terms of stock, its increase, depletion, use, conservation and its diversification. Measurements of value added are important for the organization of an industrially productive system, which is an important subsystem of the economy as a whole. But is only partially relevant to the business of measuring, targeting, and organizing the wealth of nations.

If the growth of GNP is three percent, but the uncalculated costs of output are some four percent of GNP, then at least we know that the quality of life in that country is declining. To argue that these costs be discounted is to argue for introduction of a system we term SNI, Sustainable National Income, a national income in which interest and yields are indeed added up, but in which depletion of resources and nature are subtracted from the income, as it were. Even then, the problem remains that even a corrected GNP still says nothing about the real value and dignity of a society. However (so say the politicians) without a growing GNP the country will become a second-rate nation, and so we must subject ourselves to interventions that are progressively demolishing the whole postwar social fabric. We are being colonized by the economy, as it were, and that was certainly not the original aim.

In his dialogue on wealth and welfare, Giarini points at the many paradoxical reasonings in the theory of wealth accumulation. Classical economists, and in particular Ricardo, were well aware that the methods for the accounting of economic wealth that they were devising were not really comprehensive of the real level of wealth of an individual or a country. A clear distinction was made between the notion of riches on the one hand and of wealth on the other. There was even an implicit acceptance that there could be situations where an increase in wealth would not correspond to an increase in riches.

However, these considerations remained secondary because the main problem during the Industrial Revolution was to identify the most dynamic system for increasing the wealth of nations, i.e. the industrialization process, and to concentrate on its development. Inconvenient discrepancies between wealth and riches were considered of minor importance. The writings of classical economists and of some of their later commentators were very much influenced by the fact that the first formulation of economic theory was a description of the industrialization process: the priority, which was quite adequate for this purpose, was to measure a flow of goods and the value added, whether supply-, or demand-based.

In the Service Economy, where the industrialization process per se is no longer identified as the prime mover in increasing the wealth of nations, the problem is quite different and the contradiction between wealth and riches becomes much more important.

The divergence of the notion of riches from the notion of wealth corresponds to what can be called the development of deducted values in the modern economy. Increase in these deducted values stems from the increasingly higher allocation of economic resources to activities which do not add to the real level of wealth (or of riches), but which are in fact absorbed by rising costs of the functioning of the economic system.

Let us take an example. In many households, the level of wealth is sharply increased by the introduction of washing machines, other electrical appliances, and new tools that make housework easier. But with the increased level of wealth comes an increase in the amount of waste produced in the home, which, during the 1960s, led the research divisions of companies producing household appliances to develop new machines for getting rid of kitchen waste. In a traditional sense, a waste shredder (or a waste compactor) machine adds to wealth, whereas in reality, it is merely coming with the increased nuisance at one place in the system (the private house) and creating a system breakdown elsewhere (at the sewage or waste-treatment plant). In addition, we have not become richer by having a machine to destroy garbage, as compared to when we had no garbage to get rid of. But, according to the economics of the Industrial Revolution, our wealth has increased.

Examples of this trend which began in the 1960s abound. Air and water pollution are obvious cases of diminishing real wealth (or of diminishing riches). If money is invested to de-pollute water or to develop alternative solutions such as bottled water, special reservoirs for drinking water, or swimming pools next to a polluted seashore, we are once again confronted by Catch-22 situations where investments are necessary to compensate for riches lost through, for example, pollution: these investments are not net added value to our wealth!

The growing discrepancies between levels of wealth and riches (or the contradiction between economically accounted wealth and real wealth) clearly indicate the need to refer increasingly to stock, i.e. variations in real wealth, as a substitute for the measurement of production flows (the bathtub example). Furthermore, there is also a problem of matching real added values to deducted values. A new conceptual approach to systems for measuring the real results will have to replace the simple analysis of the costs of an isolated activity.

The notion of deducted value implies the need to take into consideration the notion of negative value. In terms of economic analysis, this is already a step in the right direction, given that in many cases the negative side of economic activities has simply remained unaccounted for. Diminishing increase in an economic situation has in fact to be distinguished from a net negative process. Measuring wealth through flows that do not fill a bathtub, or even worse, that are shut off, excludes the notion of negative flows. Only by looking at the stock can positive and negative variations be measured and a decision taken as to whether the flows produce values added or values deducted.

Besides the formal conversion of the GNP into an SNI, the meta-message of this report concerns the necessity of defining economics in a final tuning way, by pointing at not only the paradoxes but also at its diversity. Economics is not a law of nature, and when it comes to output, income growth and distribution, resource use and welfare development, any system can be chosen and molded because what are involved are primarily questions of culture, choices that are made and implemented by human beings, with the economy merely a tool to help us, nothing more. Economics should then be—and can be—an instrument to define the truth.

WHAT HAPPENS IF WE FAIL IN THIS QUEST?

In the first place, we would reiterate our original message; in the words of Jay Forrester: "Over the last hundred years, life on earth was dominated by growth. Growth of population, of production, of income and capital formation, of exhaustion and pollution. This growth is going to stop and must stop, and the only question is by what means? Voluntarily, by government and free will, or through natural processes, which means collapse and disaster?"

Ultimately, this is the vision of the future, and many elements of it have already become reality in the world around us: collapse of life-support systems, of communities, regions and nations, lack of food, scarcity of water, climate change and, ultimately, war. Of the approximately 100 wars now being fought in the world, more than 70 percent originate in part in exhausted resources and collapsing life-support systems. This is the ultimate consequence, clearly confirmed by such authors as Meadows, Kennedy, Kaplan, and others (see References).

The second consequence consists in the mild precursor of this collapse, the process of individual enrichment of the few at the cost of growing public poverty, the decline in wealth and welfare to be observed everywhere today, now methodologically confirmed by the aforementioned studies of Daly and Cobb.

It is important to hold modern Western political practice up to this light, a practice consisting of ever more austerity programs to secure the integrity of purchasing power or of individual consumption, to which political affairs are being sacrificed.

Because the dominant focus of technology is to substitute labor (a process known as productivity growth), an imbalance in income growth sets in between those sectors where productivity rises—in other words, industry—and those where it cannot; in health care, education, justice, and public administration, for example. Wage demands in these sectors cannot be absorbed by rising output, although due attempts are made by amalgamating schools, closing senior citizens' homes and hospitals, abolishing police corps, and overloading the courts. The ultimate outcome is that the modern welfare society is disappearing, to the benefit of growing private consumption and the enrichment of a small elite. The neoliberal model thus becomes the future: miserable public services, bad public transport, decrepid and unsafe inner cities, overcrowded and ever more unhygienic hospitals, impoverished senior citizens; unmotivated, poor education; neglected culture, minimization of scientific research, and environmental neglect. Every government today holds up this agenda, and it is no wonder that they are all concerned above all with cranking up production growth, in the hope that this will generate funds with which to compensate for the new poverty. That may have worked with growth in the past, but it does so no longer, because an ever greater proportion of each new round of production growth consists of negative economy: compensation and repairs, processing of waste and controlling of complexity, in other words expenditure that is taken to be income. The contemporary example par excellence is in those countries which today suffer from war, guerrillas, and dictatorship, and where the arms industry is earning masses of money and, when one day there is peace, so will the demolition companies, the clear-up gangs, the contractors, the international consultancy agencies, the whole redevelopment business. When, twenty-five or fifty years from now, the country has been redeveloped to its condition prior to 1990, no net achievement will have been made, but the growth figures will be high.

This is the fate of every economy that has exceeded the limits to growth, and this means that in those countries, monetary policies are leading to accelerated demolition of both the welfare state and the cornerstones on which production growth rests.

Both forms of collapse are the result of the hypocrisy and the metaphysic bound up in the economic information. This report is about the unmasking of that hypocrisy and is thus a plea for a form of rationalization that in the world of economic metaphysics has until now proved extremely difficult. Economics can be a beautiful instrument when applied in its original meaning: to put the house (oikos) of mankind in order.

ECONOMIC GROWTH AND

WESTERN SOCIETY

CHAPTER 2

The History and Roots of Growth

 In order to understand the complexes and obsessions of individuals, it is often important to go back to their youth. The same may also hold for cultural epochs. In order to shed light on ideas about the necessity of material progress and economic growth, which without exaggeration can be called the two central obsessions of Modernity, it is useful to look at their origins. In the cultural heritage of the West we find a standard account of these origins. It tells us that the beginning of Modernity, with its dreams of progress and growth, spelled out a new awakening for mankind. When we refer to the beginning of this period as The Renaissance, or to a later phase as The

Enlightenment, this standard view is forcefully present in the names of these periods.

With the rise of post-modernism this standard account is coming under growing attack. In philosophy and the humanities the dark side of the beginnings of Modernity are drawing ever more attention. The confident and optimistic story of Modernity is slowly being unravelled and deconstructed, dreams are turning out to be nightmares, or at least attempts to keep nightmares out of sight, and promises are being unmasked as desperate responses to fears and anxieties instead of being simple beacons for a brighter future.

One of the main critiques of Modernity along these lines has been *Cosmopolis*, by Stephen Toulmin. In this study, the English philosopher undertakes to lay bare the hidden agenda of Modernity. In his prologue Toulmin invites us to back into the next millennium. The problems facing mankind at the end of the second millennium are warning us that the way straight ahead is no longer feasible. The old battle cries of Modernity no longer suffice. Rather, we should look back at the way the modern project was conceived and at the philosophical, social, and economic assumptions on which it rested. In this chapter we undertake this task partially by looking at the idea of economic growth as an answer to anxiety and scarcity.

ANCIENT FEARS

Toulmin's main critique of the standard account is that it is historically untenable. He shows *in extenso* that the period during which the first ideas of the Enlightenment originated was not accompanied by the quiet and optimistic mental climate that is often assumed. The first half of the seventeenth century should rather be seen as being among the most uncomfortable, and even frantic, periods in the whole of European history. It was a time of turmoil and crisis and the ideas of Modernity that emerged at that time can be seen as answers to that severe crisis. Certainly, those answers were effective; but, according to Toulmin, they also masked some of the essential characteristics of that crisis. Three hundred years later we are facing partly the same crisis once more, and this time we cannot run away from it by falling back on the old answers, even though we are still fascinated by them.

We find the same analysis that Toulmin expounds for the seventeenth century in the work of many historians who treat the whole period of the beginning of Modernity. As an illustration we mention *La peur en Occident*, a celebrated study by the French historian Jean Delumeau. He characterizes the transition from the Middle Ages to modern times as an era of many and major perturbations. From the middle of the fourteenth century, there was a multiplication of fears all over Europe. Fear of the devil, fear of women, of Jews, strangers, nature, and death are some of the themes treated by Delumeau. Compared with earlier and later periods, in that era there was a respectable proliferation of fright of all shapes and sizes. In this context, completely new enemies were invented, exemplified not only by the witch-hunts and the persecution of Jews, but also by the idea that nature is the enemy of man, to be defeated and subdued, an idea that dates back to the beginning of the modern era.

According to Delumeau, the many anxieties of the late Middle Ages—also described by the Dutch historian Huizinga in his famous work, *The Waning of the Middle Ages*—can be attributed largely to the loss of traditional structures and attitudes. The moral vacuum at the end of the Middle Ages made way for fear. The traditional answers to major existential questions were no longer taken for granted and lost their legitimacy as culture failed to provide new responses.

Delumeau's approach shows that the belief in progress served largely as compensation for the veiled fears of the era. The beautiful dream of better times is just one side of the coin. The optimism served equally as a mechanism for exorcising fright. Western culture has been driven not only by the allusion to a continuously improving quality of life, but also by repressed fears.

In the great philosophical systems of the seventeenth century the social reality of fear is clearly reflected. Most of the fears and anxieties of that era can be summarized under the concept of scarcity. In the middle of the century, Thomas Hobbes, who called himself a child of fear, starts considering the natural condition of mankind as one of continual fear in which the life of man is "solitary, poor, nasty, brutish, and short." And at the end of the century John Locke, who is often considered to be the grandfather of our modern economy, is also obsessed by the idea of a permanent scarcity with which mankind is faced. In contrast to Hobbes, however, he suggests a way out of this situation by formulating new ideas on growth, progress, and expansion; concepts which had never really existed before. In order to understand the force and the appeal of these ideas it is necessary to turn to the analyses of these two philosophers.

HOBBES: POWER AS A COMPARATIVE PHENOMENON

The philosophical articulation of the reality of scarcity, a phenomenon that was largely unknown before Modernity, started in the seventeenth century with the ideas of Thomas Hobbes. The best approach to the subject can be found in the famous theory of power developed by Hobbes in his major study, *Leviathan*. According to Hobbes, power manifests itself in all kinds of social institutions and structures. Like Michel Foucault in our century, he already stated that every social phenomenon is based on power relations between individuals and groups. For Hobbes, power is a comparative phenomenon. It originates from a comparison between different individual people. The goods, the objects one seeks to obtain, are not pursued for their own sake, but because someone else is also keen to obtain them. When Hobbes talks about covetousness, the desire of riches, he remarks that this name is always used in signification of blame. The reason for this is because men contending for them (riches), are displeased with another attaining them. The causal connection Hobbes is suggesting here is based on comparison. It is clear that people do not desire riches because they want them directly for themselves. Riches are important only in comparison with what others have. The possibility of being content with the riches one possesses seems to be excluded, for otherwise it is difficult to understand the fact that if others are rich, it is a source of displeasure.

Starting from this triangular comparative principle, society for Hobbes becomes a large arena of conflicts. Competition for riches, honour, command, or other power inclines toward to contention, enmity and war: because the way of one competitor to the attaining of his desire is to kill, subdue, supplant, or repel the other. That society is a zero-sum game for Hobbes is evident. Everything one attains is realized at someone else's expense. The power that people desire is always power over others, power at the expense of others. Or to state it finally in terms of the concept we are exploring here, because of the comparative relations between people, everything in Hobbes' universe potentially becomes a scarce good. His definition of power makes it impossible for there ever to be enough of it.

According to Hobbes, scarcity, the relation between limited means and unlimited ends, is caused by the continuous comparisons and strife between individuals and groups. The ends are unlimited because one must always surpass the ends of others. As a philosopher, Hobbes reflected on a new kind of social relations he saw emerging in his time. Most modern scholars agree that in the work of Hobbes we see a radical break with his philosophical predecessors. He turned the world of Plato, Aristotle, and the medieval thinkers upside down, a radical break that is best reflected in the emergence of the theme of scarcity. Before the rise of modern economic society, no one had suggested that unlimited desire was a natural quality of man. Scarcity arising out of this limitless, triangular desire is in this general sense an invention of Modernity. This is not to deny that the European Middle Ages, in common with non-Western cultures, experienced many periods of shortage and insufficiency. Until the nineteenth century, though, even in Western Europe, the noun "scarcity" generally expressed an episode of shortage, a period of insufficiency, a dearth. Only at the end of the nineteenth century did the concept of scarcity start to signify a general condition of mankind. This modern usage of the term is already foreshadowed in the work of Hobbes, who reflected and conceptualized the radical rupture in the history of mankind that consisted in the emergence of Modernity.

Society for Hobbes becomes a lifeboat in which all the passengers are fighting one another, not a culture as we know it traditionally. Survival, not the good, the right life, becomes the ultimate value. The fear of scarcity always compels one to be the first to strike or throw out the other. The relations between individuals and groups that Hobbes describes as war and that we conceptualize here as scarcity, are characterized by feelings of fear, competition, and envy. Fear is especially predominant for Hobbes. Fear largely regulates human behaviour. In order to escape fear, men by covenant create the great Leviathan, a semi-absolute state that keeps its subjects in awe and prevents the permanent scarcity from developing into outright war.

LOCKE: GROWTH AND EXPANSION

By the end of the seventeenth century, modern images of nature and the idea of unlimited progress and growth were being described by John Locke, who was aware of the threat of scarcity, describing it as a state of war. But since he did not accept that scarcity originates in human social relations, he could suggest a way out. His view of scarcity became the accepted view of Modernity, which is that scarcity is an economic, a quasi-natural fact, a perpetual relation between mankind and nature. For him scarcity simply spells out the fact that earth and nature fail to provide enough for all. Therefore we only have to produce more to alleviate scarcity or even to put an end to it. Economic growth and expansion are the answers to the threat of scarcity. With Locke, mankind, that is Western mankind, starts a rush forward in order to escape scarcity.

In this rush forward, nature—which, however mechanical a philosopher he may have been, was still seen by Hobbes as our common mother—becomes the prime enemy of mankind. Nature does not provide enough and man has to struggle and labor with her, to subdue the earth, in order to produce more and more. In the many comparisons of Locke in his Second Treatise, nature loses its traditional connotations, it is devalued at the expense of labor. Locke wonders what the difference is between an acre of land planted with tobacco or sugar and an acre of the same land lying fallow without any husbandry. When the reader makes this comparison he will find out that of all the things useful to the life of man, when he divides what in them is purely owing to nature and what to labor, he shall find that in most of them ninety-nine hundredths were wholly to be put on the account of labor.

In Locke's philosophy, the themes of progress and of the devaluation of nature are closely linked with the fear of scarcity. The European Enlightenment and the myth of progress can thus be partly interpreted as a flight forward away from scarcity.

The same can be said of European expansion. For Locke, America was an empty continent that could help alleviate the effects of scarcity in Europe. Colonialism and expansion find their legitimacy in the quest to alleviate scarcity. Cecil Rhodes, the greatest imperialist of the last century, openly stated that imperialism was necessary to avoid a class war in England. And he already dreamt about the necessity of an ongoing expansion to the universe. "I would annex the planets if I could," he wrote. To give one more example, former president Reagan in his speech at the launching of the *Discovery,* after the failure of the Challenger, told the American people that we have to conquer space in order to overcome war, scarcity, and misery on earth. The argument is exactly the same as that given by Locke in the seventeenth century.

Unlike Locke, Hobbes never saw the flight forward into empty space as a possible way to escape scarcity. In Chapter 24 of *Leviathan* he writes extensively about plantations or colonies, but nowhere does he suggest that the perpetual war among men can be solved by expansion. He knew that scarcity originates in human relations and that people trying to escape it will inadvertently spread and propagate it to the ends of the earth and even into outer space.

THE ORIGINAL MYTH OF THE MODERN ECONOMY

In comparing Europeans, who have subdued nature by way of rational labor, with native tribes such as the American Indians, Locke wrote that the latter are rich in land, but poor in all the comforts of life, because they do not work. They have the same fertile land, yet, for want of improving it by labor, have not one-hundredth part of the conveniences we enjoy, and a king of a large and fruitful territory there feeds, lodges, and is clad, worse than a day laborer in England.

   In this kind of comparison we can already discern the original myth of the modern economy. According to this myth (which was to be elaborated fully later in the eighteenth century by Adam Smith) scarcity is the original condition of all humankind, a condition with which humanity must wrestle, first in order to survive, later to live in comfort. In this myth, the typically modern human stance towards a nature that is scarce and hostile is projected back onto the whole history of mankind. Nature becomes the scapegoat of Modernity, scarcity becomes the source of violence. It is not Hobbes' competitive desires that create scarcity; it is in nature that the very causes of scarcity lie, for that is where the resources lie. If mankind can subdue and conquer nature by labor, then peace and abundance will be his future lot.

   It is important to appreciate the double role that money plays in Locke's principles of scarcity. On the one hand it is the invention of money that causes scarcity. On the other hand money is constantly luring men forward, and as such holds out the promise of helping to overcome scarcity. Originally the world provided enough for everyone. It was solely the usefulness of things that counted and it was senseless to appropriate more of the earth's products than one could use. The rule of propriety, according to Locke, was that every man should have as much as he could make use of. And for Locke it is clear that this rule would still hold in his time were it not for the invention of money and the tacit agreement of men to put a value on it, introduced by consent to larger possessions and a right to them. When men decided that a little piece of yellow metal which would keep without wasting or decay, should be worth a great piece of flesh or a whole leap of corn, the intrinsic value of things which depends only on their usefulness to the life of man was altered because now the desire of having more than man needed found a ready outlet. In this context Locke wrote the famous phrase: "Thus in the beginning, all the world was America," in which his earlier expounded ideas about how the plenty of that continent can overcome scarcity become manifest. However, he continued: "for no such thing as money was anywhere known. Find out something that has the use and value of money . . . and the gathering of possessions will start, along with an attendant increase in scarcity."

On the one hand, then, money is the cause of scarcity, by enabling everything to be translated into common terms. On the other hand money promises an end to scarcity and as such is the driving force behind the laboring of mankind. Mankind keeps on running forward because around the next corner a future of eternal abundance may always be lying in store. In the economic philosophy of Locke, scarcity is already the basic assumption underlying the economy, and economic growth is the central response to this assumption. The war of every man against every man can be averted by growth, it can be replaced by the economic war of mankind against nature that in the end will bring peace, prosperity, and abundance. The promise is clear; but behind the promise, the hidden fears stand perhaps even more prominent. Every time the promise is called into doubt, the fears make it all but impossible to question it in a rationally articulated fashion. From the very start, rationality has never held a prominent place in the limits-to-growth debate. In the final count, belief in the promises of growth and expansion is rooted in a hidden fear of scarcity, and rational debate hardly suffices to bring this fear out into the open, let alone tackle it.

TWO ECONOMIC POSITIONS

The inadequacy of rational analysis for grappling with the promise of growth comes out most clearly in some famous passages of what is perhaps the most important economic study of the 19th century. In his *Principles of Political Economy*, John Stuart Mill gives extensive space to a treatment of the necessity of a stationary economy. These passages were already highlighted in *The Limits to Growth*, the first report for the Club of Rome, but it is worth recalling them here in another light. In that report Mill's words were used as a rational exhortation. Here we would rather quote them to demonstrate the inadequacy of a rational critique of the growth credo, be it made in 1848 or 1972. All Mill's assumptions and ideas in these passages are reiterated in today's theorizing about the steady-state economy and the necessity of setting limits to growth. In a beautiful way, Mill succeeds in summarizing the essential arguments that can still be found today in the debate on economic growth and how to escape from its intrinsic dangers. For this reason it is important to quote him extensively. The failure of Mill, perhaps the most influential economic thinker of the last century, to persuade his contemporaries can serve as a warning in contemporary discussions. Only by broadening the discussion, by raising it above the narrow "rational" economic context, do we have a chance of exposing and diagnosing our society's obsession with economic growth.

> It must always have been seen, by political economists, more or less distinctly, that the increase in wealth is not boundless: that at the end of what they term the progressive state lies the stationary state, that all progress in wealth is but a postponement of this, and that each step in advance is an approach to it . . . if we have not reached it long ago, it is because the goal itself flies before us.
>
> I cannot . . . regard the stationary state of capital and wealth with the unaffected aversion so generally manifested towards it by political economists of the old school. I am inclined to believe that it would be, on the whole, a very considerable improvement on our present condition. I confess I am not charmed with the ideal of life held out by those who think that the normal state of human beings is that of struggling to get on; that the trampling, crushing, elbowing, and treading on each other's heels which form the existing type of social life, are the most desirable lot of human kind, or anything but the disagreeable symptoms of one of the phases of industrial progress . . .

The predicament described by Mill clearly resembles the scarcity that Hobbes and Locke were acknowledging in their century. People still were struggling to get on in order to reach the final goal of prosperity. Mill endeavours to show in a rational way that the goal is flying before them. He attacks the mere increase of production and accumulation as a lie. I know not why it should be a matter for congratulation that persons who are already richer than anyone needs to be, should have doubled their means of consuming things which give little or no pleasure except as a representation of wealth. Mill's plea for a steady state also runs counter to the idea that humanity must subdue nature in order to hasten the growth of production. Nor is there much satisfaction in contemplating the world with nothing left to the spontaneous activity of nature; with every rood of land brought into cultivation, which is capable of growing food for human beings; every flowery waste or natural pasture plowed up, all quadrupeds or birds which are not domesticated for man's use exterminated as his rivals for food, every hedgerow or superfluous tree rooted out, and scarcely a place left where a wild shrub or flower could grow without being eradicated as a weed in the name of improved agriculture. Unlimited increase of wealth, according to Mill, devastates the earth and makes it lose its pleasantness.

Finally, Mill tries to convince his contemporaries that in a steady state they would be far better off than in the present rush forward. It is scarcely necessary to remark that a stationary condition of capital and population implies no stationary state of human improvement. There would be as much scope as ever for all kinds of mental culture, and moral and social progress; as much room for improving the Art of Living and much more likelihood of its being improved, when minds cease to be engrossed by the art of getting on. Even the industrial arts might be as earnestly and as successfully cultivated, with this sole difference, that instead of serving no purpose but the increase of wealth, industrial movements would produce their legitimate effect, that of abridging labor.

All these arguments of Mill seem rational enough. But they did not touch his contemporaries, exactly because they were only rational. Mill did not detect the hidden fears behind the struggling to get on, he did not notice how a fear of scarcity in a certain sense compelled his contemporaries to continue down the well-trodden path. Because of this blind spot in his analysis, neither could he experience the force and deep appeal of the promises of growth. For him these were merely irrational and he considered it his duty to demonstrate their irrationality. This promise was not irrational at all, however, for, crucially, it offered a strategy for avoiding repressed fears and sentiments.

In our century, economists hardly reflect on the assumptions of their profession any more. They do not talk about fears and promises, they play the in-between, providing their models and calculations. When they give free rein to philosophical reflection, however, the old themes reappear. And in the work of a leading 20th century economist, the old promises gain new force, as he is so bold as to even mention a date for their fulfillment.

In 1930, John Maynard Keynes prophesied that within two generations industrial societies might finally realize what in his view had always been the goal of humankind: an end to the problem of scarcity. Keynes, seeing his contemporaries still struggling down the dark tunnel of scarcity, perceived a future that would bring light in the form of satisfaction of all basic needs.

> When that day came, a change in moral codes would also be possible. We shall be able to rid ourselves of many of the pseudo-moral principles which have hag-ridden us for two hundred years, by which we have exalted some of the most distasteful of human qualities into the position of the highest virtues . . . The love of money as a possession . . . will be recognised for what it is, a somewhat disgusting morbidity, one of these semi-criminal, semi-pathological propensities which one hands over with a shudder to the specialists in mental disease. When scarcity is overcome we shall be able to return to the traditional virtues of mankind. We shall honour those who can teach us how to pluck the hour and the day virtuously and well.

These well known phrases of Keynes have important implications that remained hidden for their author. First, Keynes suggests that the tunnel of scarcity has a beginning. Before that beginning, somewhere in the past, virtuous living was apparently possible. With the coming of the modern economy, however, it seems that the world was turned upside down. The promise of the economy is that in the future it will again be possible to live humanely and virtuously. But Keynes warns his contemporaries that before that time comes, they should continue to live in their upside down world of scarcity.

> Maybe for another hundred years we shall have to tell ourselves and each other that honesty is mean and meanness is honest. Meanness is useful and necessary and honesty is not. Only the characteristics of meanness, envy, greed and competition can lead us ultimately out of the dark tunnel of scarcity. Starting from a Hobbesian analysis it becomes clear that the tunnel of scarcity rather resembles some kind of cyclone in which men become caught, spinning around in rivalry with one another and in the process creating more and more needs and desires.

The two generations as well as the greater part of the hundred years speculated on by Keynes have passed, and scarcity is becoming an ever greater threat. It is one of the key words in the analysis of *Our Common Future*, the report of the Brundtland Commission, and it is a central theme in every analysis of our environmental problems. But still the answer seems to be the same as Locke's: economic growth. In order to understand why it is exactly this economic growth that is now creating scarcity rather than alleviating it, we must take this historical elucidation to its final step.

SCARCITY AS A SOCIAL CONSTRUCTION

In a sense, the concept of scarcity was more or less invented in the seventeenth century; however, the term "invention" might suggest that at a given moment someone just thought up and propagated scarcity. It is more likely, however, that the modern economy autonomously created the idea of an ever-present universal scarcity as the founding myth of modern society. The rise of scarcity is better explained on the basis of the concept of social construction. Hobbes and Locke looked at their society and saw that something like scarcity was present in many human and social relations. In his *Second Treatise*, Locke systematized the way the upper classes in England were already trying to escape from scarcity, but in the long run, this presumed escape only reinforced scarcity and made it omnipresent. By pursuing ever more production, the whole of culture and nature was thus reduced to a field of scarce resources. Viewing the world in this way, and above all acting on this view, Western society has succeeded in establishing the regime of scarcity in almost all spheres of social and personal life, worldwide.

To put it bluntly, by acting for several centuries upon the assumption that nature is just dead and worthless stuff that has to be managed by human labor and technology, we have almost succeeded in making it dead and worthless. By defining everything as a scarce resource and acting this out, we are in the long run making it so. Hobbes' analysis finally proves to be true. It is impossible to escape scarcity by growth and expansion. For three centuries, Locke's solutions have carried great weight. As the twentieth century closes, Hobbes' diagnosis of scarcity is evidently more fundamental. We do not have to adopt the great Leviathan as a therapy for the problem of scarcity to acknowledge the depth of the diagnosis. Hobbes was not ashamed to label himself a child of fear. We, in turn, should show no hesitation in uncovering the hidden historical (and still extant) fears behind the promises of economic growth. Only then will we be able to pose the crucial question: will economic growth, which is still seen as the main remedy for alleviating scarcity, in fact propagate scarcity all over the world, to such an extent that finally this world will come to an end, "not with a bang but a whimper?" (T.S. Eliot).

CHAPTER 3

Economic Growth and Gross National Product

ORIGINS

The first studies about national income stemmed from curiosity about the differences in economic strength between nations. In seventeenth-century Britain, Petty and King made comparisons between Britain, France, and Holland (republished: King, 1936. Similar work was done in France, but for a long time there was no follow-up in either country. Around the beginning of the nineteenth century, interest in the subject revived. Metelerkamp (1804) visited several European countries and found fairly recent estimates for France, Britain, the Netherlands, and Saxony. This was followed by another long period of silence.

The twentieth century is the age of modelling and quantifying national economies. Generally speaking, the interest in national income studies is stimulated by several circumstances. First, the political willingness to steer economic development is important, certainly when social consequences have to be taken into account politically. Second, there has to be a theoretical framework, since insight in the coherence of economic processes leads to an understanding of relationships. And, third, the sheer availability of quantitative data on the economic process will create an interest in what we now call national accounting.

The political will came into play in the thirties. The Great Depression, the unemployment rate, and—given the experience of economic problems in the First World War—the threat of another war confronted governments with problems that, although not new, were on a scale hitherto never encountered. The theory was developed after the First World War Clark, Kuznetz, and Leontief were active in the United States. In Britain, Keynes, Meade, and Stone played their part. In Norway, contributions were made by Frisch and Aukrust, and in the Netherlands by Tinbergen and Derksen. Then quantification followed. As early as 1920, there was an input-output table for the Russian economy, but this table was not used to base economic policy on. The Americans developed their quantitative data in the thirties, and the English published their 'White Paper' (entitled "An Analysis of the Sources of War Finance and an Estimate of the National Income and Expenditure in 1938 and 1940") during the Second World War. In Norway and the Netherlands, the first quantitative data appeared in the thirties.

There were three related aspects in this development. First, researchers realized that information on the national income alone could not provide answers to the problems posed. Consequently, measurement was extended to related totals such as household consumption and fixed capital formation, which were eventually incorporated with the national income in a system of accounts that initially described a large part of the economy and later all of it. Second, since existing statistics covered only part of the economic process and did not always fit the system even when they were available, new methods of estimation for filling in the gaps had to be developed. Finally, there was econometrics, making efforts not only to measure, but also to

NOTE: The authors want to thank Carsten Stahmer for his helpful comments.

explain economic trends. The Economic Intelligence Service of the League of Nations—the predecessor to the United Nations—set up a research project in which Von Haberler and Tinbergen participated.

All this research led to a better understanding of economic processes and of economic and social problems. As Robinson, who worked with Keynes at Cambridge, wrote in 1986: "Economics was never the same again." He could have added: and neither was the role of government.

THE SYSTEM

The national accounts provide a quantitative description of the economic process in a given country. The word "quantitative" implies the step from economic concepts to statistical measurement. Many detailed definitions and descriptions have been developed in the course of many years. Measurements require these details when the statistical tables are to be used for economic and social policies, and even more so when they have to be internationally comparable. This requirement of detail, however, masks the fact that the system is basically very simple.

Early researchers took the bookkeeping system of enterprises as their model (Table 3.1). This system shows the assets and liabilities of the enterprise and all its transactions with others. Balance sheets and profit-and-loss accounts are routinely produced from this registration, but the directors of the firm can ask for the basic information to be regrouped if this is required by the problems they face.

Instead of registering the transactions of one unit with all others, the national accounts came to register transactions between all the enterprises, institutions, and individuals in a given country. Because there are millions of these "actors" in the economy, a clear, overall picture requires a classification of the actors into broad sectors, distinguished by their function in the economy. Next, the transactions between sectors must be classified into groups that serve the description and analysis of the process. These transactions will be entered into accounts—if necessary, more than one for each sector. Finally, transactions have to appear twice in the system, because they are between sectors.

During the Second World War, work on such schemes was undertaken in several countries quite independently. These early studies

---

Table 3.1

EXAMPLE OF BOOKKEEPING SYSTEM

1.1 Purchases of goods and services	2.1 Sales of goods and services from other producers to other producers
1.2 Purchases of imported intermediary goods and services	
1.3 Compensation of employees	2.2 Sales of consumer goods to households
1.4 Entrepreneurial income	
1.5 Indirect taxes minus subsidies	2.3 Sales of consumer goods to the government
1.6 Depreciation allowances	
	2.4 Sales of investment goods to enterprises and the government
	2.5 Increase in stocks
	2.6 Exports
TOTAL	TOTAL

---

concentrated mainly on the processes of production, income formation, and spending. This usually resulted in production accounts for the sectors enterprises and government, an income and consumption account for households and for the government, an account for fixed capital formation of all sectors together, and an account for the transactions with the rest of the world. There were some variations; and such differences can be quite acceptable if the wish for international comparability is dropped.

Setting up the list of transactions produced hardly any problems: there is little freedom of choice. Choices were necessary for fixing the boundaries of the system. The possibility to measure transactions and especially to measure the value of these transactions meant the system had to be restricted in principle to transactions which generate a measurable money flow. Work in the households or volunteer work was therefore excluded from the system. Later, accounts were added about the distribution and redistribution of income and for the financing transactions of capital formation. But the original scope of the system still forms its core, and that is the key to understanding it.

For the producers in the economy (mostly enterprises, but other institutions and the government as well), the information about the transactions that take place in their part of the production process can be presented as in Table 3.2. Tables 3.2-3.5 can be seen as aggregated results of the input-output table, Table 3.6, which will be discussed below.

The two totals in Table 3.2 must be equal because the same transactions are involved. Table 3.2 is not quite complete yet. The total of imports can be added as item 1.3 in Table 3.3, while the value of the imported consumer goods and investment goods separately can be added to items 2.2-2.6 in this table.

If this table is constructed for all producers and the results are added, items 1.1 and 2.1 are equal (36,768), so they can be left out. This is done in Table 3.4. This result is a new table, referred to as the "primary inputs and final use of products account." The left-hand side shows the goods and services that are available in the economic process of a nation in a given period. The first four items together form the Gross Domestic Product (GDP) as an aggregate of the goods and services produced on the nations' territory. The item Imports stands for the goods and services produced abroad that have become available in the economic process. The right-hand side of Table 3.4 shows the final destination of the available goods and services. Of course, the destination often depends on the nature of the commodities.

Table 3.2

DOMESTIC PRODUCTION ACCOUNT

1.1 Intermediate consumption of domestic products	24,789	2.1 Intermediate consumption of domestic products	24,789
1.2 Intermediate consumption of imported products	11,979	2.2 Household final consumption of domestic products	17,854
1.3 Gross value added	32,292	2.3 Government final consumption of domestic products	4,594
• Compensation of employees	14,761	2.4 Gross fixed capital formation of domestic products	6,263
• Entrepreneurial income	11,454	2.5 Increase in stocks of domestic products	485
• Indirect taxes minus subsidies	3,094	2.6 Export of domestic products	15,075
• Depreciation allowance	2,983		
GROSS OUTPUT	69,060	GROSS OUTPUT	69,060

Table 3.3

SUPPLY AND DISPOSITION OF PRODUCTS ACCOUNT

1.1	Intermediate consumption of products	36,768	2.1 Intermediate consumption of products	36,768
1.2	Gross value added	32,292	2.2 Household final consumption	19,537
	Gross output	69,060		
1.3	Imports of products	16,443	2.3 Government final consumption	4,913
	• Intermediate products	11,979		
	• Final products	4,464	2.4 Gross fixed capital formation	8,119
	• Household consumption	1,683	2.5 Increase in stocks	723
	• Government consumption	319	2.6 Exports of products	15,443
	• Gross fixed capital formation	1,856		
	• Stocks	238		
	• Exports	368		
	SUPPLY OF PRODUCTS	85,503	DISPOSITION OF PRODUCTS	85,503

Table 3.4

PRIMARY INPUTS AND FINAL USE OF PRODUCTS ACCOUNT

Compensation of employees	14,761	Household consumption expenditure	19,537
Entrepreneurial income	11,454		
Indirect taxes minus subsidies	3,094	Government consumption expenditure	4,913
Depreciation allowances	2,983	Gross fixed capital formation	8,119
Gross domestic product (GDP)	32,292	Increase in stocks	723
Imports of goods and services	16,443	National final expenditure	33,292
		Exports of goods and services	15,443
TOTAL	48,735	TOTAL	48,735

Table 3.4, end result of a long process of economic research and measurement, is central to the discussions of plans for government policies. The debate on the role of government and on economic theory continues, of course, and the exact definitions of economic variables depend on the state of the economic theory and technology. But Table 3.4 describes an identity. Goods and services have their origin and their destination in the economic process. Therefore, the system of definitions of economic variables have to be coordinated. After all, it is exceedingly difficult to consume a commodity that has not been produced or imported.

Other totals can be derived from the gross domestic product at market prices:

- by adding net factor income from abroad: gross national product at market prices, which (at current prices) equals gross national income at market prices;
- by deducting depreciation allowances (= consumption of fixed capital): net national product at market prices;
- by further deducting indirect taxes minus subsidies: net national product at factor costs.

Table 3.5 (right-hand side) gives a summary of these macroeconomic totals.

The totals are required for different purposes. Factor costs are costs of the production factors, otherwise known as Labor and Capital. A study focusing on gross earnings (or primary income) of production factors, for example, has to eliminate the indirect taxes and subsidies that influence the marketprice. Also, a gross aggregate does not take into account the losses in capital stock as a consequence of the production process. And, there is a difference between an income and a production concept in a open economy, because of possible differences in price movements of imported goods and services and exported goods and services. This means that the "value" of income (the packet of goods and services that can be bought with a given income) generated by the production process may differ.

All these product totals can be readily expressed in constant prices. The starting point for the national income is the net national product at factor cost in constant prices. But a correction must be made for the fact that some goods forming the net national product packet can be imported and exported. If export prices go down rel-

Table 3.5

DOMESTIC AND NATIONAL PRODUCT

Gross domestic product		Gross domestic product	
(GDP)	32,292	(GDP)	32,292
		+ (primary) income from the rest of the world	423
		− (primary) income to the rest of the world	217
		Gross national product (GNP) at market prices	32,498
− consumption of fixed capital	2,983	− consumption of fixed capital	2,983
Net domestic product (= Net value added)	29,309	Net national product (NNP) at market prices	29,515
		− indirect taxes minus subsidies	3,094
		Net national product (NNP) at factor costs (= National income)	26,421

ative to import prices (a change of the terms of trade), the packet received will be smaller than if the two price trends had been equal. In constant prices, the packet that becomes available for the national economy will therefore be smaller than the packet produced. The opposite is likewise possible. The corrected total is called real national income. If this seems somewhat complex, consider the situation for a country producing oil and little else. If it doubles its production in two years and at the same time sees the price of oil halved, its net domestic product at constant prices will double, but its real national income will be unchanged since the imports that it can buy in exchange for its oil will remain at the same level.

So Tables 3.2-3.4 can be seen as the consolidated form of a more detailed presentation of the production process, the so-called input-output table. This is a double-entry table in which the information of Tables 3.2-3.4 is given by branch of activity. Table 3.6 presents a model in which the breakdown of activities is, for practical purposes,

Table 3.6

MODEL OF INPUT-OUTPUT TABLE

		Input by branch of activity				Categories of final demand						
		Agriculture, fishing	Manufacturing, mining	Construction	Services	Total, columns 1–4	Household consumption expenditure	Government consumption expenditure	Domestic fixed capital formation	Increase in stocks	Exports	Total, columns 6–10	Total, columns 5+11
		1	2	3	4	5	6	7	8	9	10	11	12
Output by branch of activity	1. Agriculture, fishing	848	3,079	—	35	3,962	745	10	—	3	1,179	1,937	5,899
	2. Manufacturing, mining	1,465	9,500	1,389	2,142	14,496	9,375	1,015	1,899	407	8,485	21,181	35,677
	3. Construction	56	206	384	407	1,053	166	256	3,597	—	35	4,054	5,107
	4. Services	240	2,218	378	2,442	5,278	7,568	3,313	767	75	5,376	17,099	22,377
	5. Total, rows 1–4	2,609	15,003	2,151	5,026	24,789	17,854	4,594	6,263	485	15,075	44,271	69,060
Primary inputs	6. Imports	196	9,022	799	1,962	11,979	1,683	319	1,856	238	368	4,464	16,443
	7. Compensation of employees	633	5,608	1,267	7,253	14,761	—	—	—	—	—	—	14,761
	8. Entrepreneurial income	2,248	3,369	665	5,172	11,454	—	—	—	—	—	—	11,454
	9. Indirect taxes minus subsidies	—	1,549	177	1,389	3,094	—	—	—	—	—	—	3,094
	10. Depreciation allowances	234	1,126	48	1,575	2,983	—	—	—	—	—	—	2,983
	11. Total, rows 6–10	3,290	20,674	2,956	17,351	44,271	1,683	319	1,856	238	368	4,464	48,735
	Total	5,899	35,677	5,107	22,377	69,060	19,537	4,913	8,119	723	15,443	48,735	48,735

limited to four branches. One of the first four rows of the table describes the destination of the goods and services produced in the branch of that row. It shows the final categories (consumption, investments, and exports) and the intermediate destinations as well. Goods or services that are the output of one branch can be the input of another. Furthermore, there are so-called internal deliveries. For example, seeds are both output and input in agriculture.

The first four columns focus on the production processes of the branches, showing how the total production of a given branch is formed. The first four rows of a column show the internal and other intermediate deliveries which, put together (on line 5), form the total input with domestic origin in the branch. The total of these domestic inputs and the inputs from abroad or imports (line 6) is the total input in the branch of activity. The difference between total output (for agriculture 5,899 in Table 3.6) and total input (2,609 plus 196 in agriculture) is the gross value added of the branch of activity. The total of gross value added in all branches of activity together (the sum of lines 7–10 in column 12) is the Gross Domestic Product.

The supply and disposition of goods and services, as stated in Table 3.4, can be derived directly from the input-output table. The row total of GDP and imports (line 6 of column 12) add up to the same amount (48,735) as the column totals for the expenditure categories. These are consumption (column 6 plus 7 on line 12), fixed capital formation (column 8, line 12), increase in stocks (column 9, line 12) and exports (column 10, line 12).

The example in Table 3.6 was limited to four branches of activities for the sake of simplicity. In real life, tables can be constructed with hundreds of branches. Furthermore, there are tables that show commodity flows explicitly: the "make-and-use" tables. Here, the basic product of statistical observation is the dimension branch × commodity. It is not only possible to derive input-output tables, and dimensioned branch × branch like Table 3.6, but derivatives with dimensions commodity × commodity are also possible. All these tables are useful for two reasons, one statistical and the other analytical.

Bringing together all the statistical information on output and input of goods and services (and the income originating in the production) in one table means that any entry in the table can be used twice: an entry in a row automatically means an entry in a column, and vice versa. This proved a great help at the time when existing statistics covered only a small part of the table and had to be supplemented by estimates. In other cases there are many possibilities of calculating missing figures as residuals, since the table covers a large number of balances in which production plus imports of goods and services are confronted with information on uses and exports.

As more statistics became available, some of these uses became superfluous, but the discipline that the table represents is still useful in the construction of the system of national accounts. The more "superfluous" the availability of statistical sources is, the more accurate the macroeconomic totals that can be calculated. Sometimes statistical results (in the form of averages, for example) can be conflicting, simply because of the nature of statistics. Also, one has to deal with nonresponse, or sample and measurement errors. When such things occur, compiling the input-output table is a way to integrate all of the statistical sources. Commodity flows, for example, have to be balanced.

New statistical uses for the input-output table were found when the demand for recent figures came up. Recent, that is, quarterly and monthly, statistics are available for many transactions of the system, but by their nature they tend to give less detail and are less accurate than the results of the pondered yearly inquiries.

Therefore, the latest input-output tables are often used in combination with recent statistics to get estimates for the most important totals of the national accounts. In the Dutch situation, input-output tables are used to make extrapolations to the most recent years and quarters. Here, we use a process of statistical integration for recent monthly and quarterly data to obtain the quarterly national accounts. Each month, much statistical information becomes available about business cycle developments. The main indicators describing the business cycle are published monthly in the business cycle report, which also outlines recent economic developments. A full integration in an input-output table on a monthly basis is virtually impossible, both for conceptual and practical reasons (Algera and Janssen, 1991).

The analytical use of the table must also be mentioned: it provides the means to measure the ratios between the primary inputs and the categories of final demand. Examples are questions about the increase of imports if household consumption expenditure increases with a certain percentage, or the direct impact of an increase of capital expenditure on the production level of the various branches of activity. The tables play a similar role in the analysis of the origin of price movements because an input-output table provides a consistent framework of the relationships between economic actors in the production process. For analytical purposes, this means that they can show not only first-order reactions to a given situation (for example, a simulation of a specific policy), but also second-, third- and higher-order consequences. A policy of building extra houses, for example, not only means more employment in construction industry, it also creates more jobs in branches of activity that produce the inputs for construction industry.

USE OF THE ACCOUNTS

The history of the application of the national accounts in government policies has so far not been written. We would like to describe a few parts of it here. The development of the national accounts in the United States and in Great Britain resulted in part from the preparation for the Second World War. The financing of this war by both countries was subsequently based on these accounts.

New ideas about the role of government in the economic process (Keynes), ideas about modelling the economic process (Tinbergen), and ideas about monitoring the economic process by a systematic, statistical framework were developed, and their impact was first felt, in international affairs. Although the process of rebuilding the shattered economies of Europe started at once after the Second World War, it ran into problems after a few years when existing reserves of foreign currencies became scarce. The United States government then launched the so-called Marshall Plan which provided financial help. Countries that applied for this assistance were required to present their plans for further reconstruction with the help of the newly named national accounts. Yearly reports on progress were also to be accompanied by tables.

In order to secure uniformity in the presentation of these reports, work on detailed guidelines for tables, concepts, and definitions was undertaken by the new Organization for European Economic Co-operation (OEEC). This was the dawning of a new era, with work on the construction of national accounts part of the regular task of government agencies in many European countries. Again in the field of international relations, the new way of economic thinking, and perhaps also the success of the Marshall Plan, resulted in plans for providing assistance to poor nations by the rich nations. Here, the international agencies involved found in the national accounts a useful instrument for assessment and planning, resulting in even more demand for the accounts.

International agencies and associations, organizing conferences and assisting in expanding the guidelines played an indispensable role in the further development of the system. This started in 1947 with "Measurement of National Income and the Construction of Social Accounts," started by the Committee of Statistical Experts of the League of Nations and finished by the United Nations which took over these activities. Subsequent work by the OEEC has already been mentioned. The fact that this was a European-American undertaking again brought the United Nations into the picture: the work of its Statistical Office on the basis of many conferences resulted in reports with revised guidelines in 1953, followed by revisions and a new issue in 1968. The latest revision appeared in 1993.

A few years after the Second World War, the International Association for Research in Income and Wealth was founded. It also made a major contribution to the stream of ideas and suggestions. In all its meetings and in the conferences organized by agencies of the United Nations the freely exchanged ideas of economists and statisticians greatly helped to improve the quality of the work. Less is known about how new tools found their place in the policies of individual countries. Statistics on the major items of the accounts are often found in budget papers, but their concrete role—if any—in the formulation of policies is not clear.

Even in the Netherlands, where the government had created a Central Planning Office as early as 1946, the acceptance of its analyses and projections took a long time. But this was understandable. In the first decade following the war, the whole theory, application of models, and, last but not least, compilation of the necessary statistical information was still in its infancy.

Nevertheless, the emergence of similar economic policies in the countries of the West shows that the introduction of the new way of thinking about economic problems, and of the role of government, took place along similar lines. There must have been a definite dividing line between the architects of the system and its builders on the one hand and politicians on the other. The transfer of ideas across that line first had to relate to the purposes to be achieved. These differed widely: on both sides of the line there were those who wanted the state to remake society and those that advocated more modest policies, such as trying to dampen the fluctuations of the business cycle or set up systems of social security for the poor.

At first, the second choice, with some variation, seemed to have been accepted in the Western world with very little discussion between governments. Later, there were discussions on the possibilities of applying the system of national accounts for other political purposes. Although it was recognized that the possibilities for creating the society were limited, the importance of economic and statistical models and other tools to explain economic events remains clear. Nowadays the relationship between economists and statisticians on the one hand and politicians on the other hand seems almost the reverse of what it was right after the war. Analysts sometimes have to warn politicians not to give an absolute interpretation to model results.

Next, the possibilities of practical application had to be explored. A factor that helped was the realization that the system of national accounts gives a coherent picture of economic development in which tendencies that would cause disequilibria, if unchecked, can be identified. A list of such tendencies (in relation to the balance of payments, the labor market, prices) led easily to an attractive list of government policies. Also, the system produced figures and in some countries projections for the GDP, and for growth, so politicians quickly learned that growth meant greater opportunities for government programmes.

FURTHER DEVELOPMENT OF THE SYSTEM

The economic and social policies of the Western countries after the war differed from earlier policies in that they dealt with the economy as a whole. For a long time these policies were quite successful in the countries themselves: incomes reached levels nobody would have dared to expect at the beginning of the period but in the 1980s and 1990s, however, growth slowed down. Economic assistance to what came to be called the Third World did not prosper: there were some bright spots, but in many cases the help given had no lasting results, and sometimes progress was swept away by population growth.

The discussion of these problems largely concerns policy changes, but there is also reference to the definitions in the SNA, specifically the definition of growth as incorporated in the GDP. Although these definitions were acceptable in the period in which they were established, it became gradually clear that they did not fit the emerging situation.

Stated simply, the problem is that the production of goods and services, as measured by the GDP, although it is an important element of the change in welfare, gives an incomplete picture: other factors must be taken into account to complete it. The international debate contained many proposals for bringing more welfare elements into the system. The problem of most of the proposals was quantification. As was stated before, the boundaries of the system were simply and clear: transactions should be measurable by market prices and/or a real flow of money in the opposite direction of the flow of goods or services. Assumptions are needed for a quantification of most of these welfare elements. This means that often (subjective) valuations start to play a role.

Van Bochove and Van Tuinen therefore proposed a system of a general-purpose core supplemented with special modules (1985). The core is a fully fledged, detailed system of national accounts, with a good international comparability. The core sytem contains the hard, quantitative, coordinated information on the economic process [production, (re-)distribution of income, income expenditure, saving and gross fixed capital formation, and the process of financing]. The modules are more analytic and reflect special purposes and specific theoretical views. The term "satellite accounting" is adopted in the international literature, in this context. In the System of National Accounts (SNA) of 1993, Chapter XXI is called "Satellite Analysis and Accounts."[1]

In the long discussions on the SNA, some key problems were presented quite early by the Scandinavians who stated that the system did not take account of the loss of their forests by extensive harvesting. At the time, although the idea was recognized as correct, it was argued that the growing of forests was not included either, that inclusion in the system would be very difficult, and that the problem was not important anyhow. Elements that might be taken into account in a more satisfactory measure of growth could be war and the threat of war, crime and corruption, the environment, depletion of natural resources, volunteer work, housework, inequality of the distribution of income and wealth, and unemployment.

One subject for which the need for better measurement is now acutely felt is the environment. As the unpleasant side effects of

---

[1] The ideas were not completely new. In an earlier paper, some headlines were set out. See Van Eck (1983).

growth in that field have multiplied, quantitative information on many kinds of pollution have been developed, as have government policies that try to stem the tide. The idea of an alternative estimate for the GDP which would account for the side effects mentioned has been around for some time. Section D of Chapter XXI in the SNA of 1993 deals with the "Satellite System for Integrated Environmental and Economic Accounting."

In the Netherlands, the so-called National Accounting Matrix, including Environmental Accounts (NAMEA), was developed (De Haan et al., 1993). In the NAMEA, environmental indicators are put on a par with the major aggregates in the national accounts, like the National Income. Extra rows and columns are added in the National Accounts matrix for information on emissions of toxic materials, supply and reuse of waste, the production and use of raw materials and energy, and, in principle, the use of land, water and air. The environmental indicators reflect the goals of the environmental policy of the Dutch government. The NAMEA is suited as a data base for modelling the interaction between the national economy and the environment. A further step will be the calculation of environmental losses in primairy costs, based upon NAMEA. This is done in a project called Sustainable National Income (SNI).

Just as the national accounts were useful when politicians developed their policies with a picture of the whole economy before them, environmental policies might become more effective as part of overall policies if the problems of the environment could be presented as part of a new and more complete picture of the economy.

PART · TWO

PARADOXES OF GROWTH

CHAPTER 4

State of the Environment

 It takes no stretch of the imagination to see that the human species is now an agent of change of geologic proportions. We literally move mountains to mine the earth's minerals, redirect rivers to build cities in the desert, torch forests to make way for crops and cattle, and alter the chemistry of the atmosphere in disposing of our wastes. At humanity's hand, the earth is undergoing a profound transformation—one with consequences difficult to grasp.

It may be the ultimate irony that in our efforts to make the earth yield more for ourselves, we are diminishing its ability to sustain life of all kinds, human included. Signs of environmental constraints are now pervasive. Cropland is scarcely expanding any more, and a good portion of existing agricultural land is losing fertility. Grasslands have been overgrazed and fisheries overharvested, limiting the amount of additional food from these sources. Water bodies have suffered extensive depletion and pollution, severely restricting future food production and urban expansion. The stability of the atmosphere has been disrupted, with heat-trapping greenhouse gases increasing and the life-protecting ozone layer diminishing. And natural forests—which help stabilize the climate, moderate water supplies, and harbor a majority of the planet's terrestrial biodiversity—continue to recede.

These trends are not altogether new. Human societies have been altering the earth since they began. But the pace and scale of degradation over the last several decades is historically new. Since 1950, three trends have contributed most directly to the excessive pressures on the earth's natural systems: the doubling of world population, a 250 percent increase of per capita economic output, and the widening gap in the distribution of income. A search for the root causes of environmental decline reveals that the environmental impact of the human population, now 5.5 billion, has been vastly multiplied by economic and social systems that strongly favor growth of production and ever-rising consumption over equity and poverty alleviation; and that do not discriminate between means of production that are environmentally sound and those that are not.

The central conundrum of sustainable development is now all too apparent: population and economies grow exponentially, but the natural resources that support them do not. In recent years, the global problems of ozone depletion and greenhouse warming have underscored the danger of overstepping the earth's ability to absorb our waste products. Less well recognized, however, are the consequences of running down the natural capital of essential renewable resources—especially land, water, and forests—and how far along that course we may already be.

THE RESOURCE BASE

Biologists often apply the concept of "carrying capacity" to questions of population pressures on an environment. *Carrying capacity* is the

EDITOR'S NOTE: This chapter is drawn from Sandra Postel, "Carrying Capacity: Earth's Bottom Line," in Lester R. Brown et al., *State of the World 1994* (New York: W.W. Norton, 1994).

largest number of any given species that a habitat can support indefinitely. When that maximum sustainable population level is surpassed, the resource base begins to decline—and sometime thereafter, so does the population.

The outer limit of the planet's carrying capacity is determined by the total amount of solar energy converted into biochemical energy through plant photosynthesis, minus the energy those plants use for their own life processes. This is called the earth's net primary productivity (NPP), and it is the basic food source for all life on earth.

Prior to human impacts, the earth's forests, grasslands, and other land-based ecosystems had the potential to produce a net total of some 150 billion tons of organic matter per year. Biologist Vitousek and his colleagues estimate, however, that humans have destroyed outright about 12 percent of the potential terrestrial NPP and now directly use or co-opt an additional 27 percent (Vitousek, 1986).

It may be tempting to infer that, at 40 percent of land-based NPP, we are still comfortably below the ultimate limit. But this is not the case. We have appropriated the 40 percent that was easiest to acquire. Although it may be impossible to double our share, theoretically that would happen in just 60 years if the human share rose in tandem with population growth. And if average resource consumption per person continues to increase, that doubling will occur much sooner.

Perhaps more important, human survival hinges on a host of environmental services provided by natural systems—from forests' regulation of the hydrological cycle to wetlands' filtering of pollutants. As we destroy, alter, or appropriate more of these natural systems for ourselves, these environmental services are compromised. At some point, the likely result is a chain reaction of environmental decline— widespread flooding and erosion brought on by deforestation, for example; or worsened drought and crop losses from desertification; or pervasive aquatic pollution and fisheries losses from wetlands destruction. The simultaneous unfolding of several such scenarios could cause unprecedented human hardship, famine, and disease. Precisely when vital thresholds will be crossed, no one can say. As Vitousek and his colleagues note, those "who believe that limits to growth are so distant as to be of no consequence for today's decisionmakers appear unaware of these biological realities" (Vitousek, 1986).

What do we do when we have claimed nearly all that we can from the earth, yet our population and demands are still growing? This is precisely the predicament we now face.

CROPLAND

Cropland area worldwide expanded by just 2 percent between 1980 and 1990, to a total of ca. 1.44 billion hectares. This means that gains in the global food harvest came almost entirely from raising yields on existing cropland. Most of the remaining area that could be used to grow crops is in Africa and Latin America; very little is in Asia. The most sizable near-term additions to the cropland base are likely to be a portion of the 76 million hectares of savanna grasslands in South America that are already accessible and potentially cultivable, as well as some portion of African rangeland and forest. These conversions, of course, may come at a high environmental price, and will push our 40-percent share of NPP even higher.[1]

Moreover, a portion of any cropland gains that do occur will be offset by losses. As economies of developing countries diversify and as cities expand to accommodate population growth and migration, land is rapidly being lost to industrial development, housing, road construction, and the like. Canadian geographer Vaclav Smil estimates, for instance, that between 1957 and 1990, China's arable land diminished by at least 35 million hectares—an area equal to all the cropland in France, Germany, Denmark, and the Netherlands combined. At China's 1990 average grain yield and consumption levels, that amount of cropland could have supported some 450 million people, about 40 percent of its population (Smil, 1992).[2]

In addition, much of the land we continue to farm is losing its inherent productivity because of unsound agricultural practices and overuse. The Global Assessment of Soil Degradation, a three-year study involving some 250 scientists, found that more than 550 million hectares are losing topsoil or undergoing other forms of degradation as a direct result of poor agricultural methods (Oldeman et al., 1993).

Environmental emissions are another cause of productivity loss. A survey of studies on air pollution damages to agriculture and forestry showed national estimates ranging from 0.2 to 7 billion US$ (Strand and Wenst p. 19).

On balance, unless crop prices rise, it appears unlikely that the

---

[1] Cropland expansion from FAO (1992); figure of 76 million from FAO (1993b), and from Thomas (1993).

[2] Cropland areas of European countries from FAO (1992); number of Chinese people that 35 million hectares could support is Worldwatch Institute estimate, based on U.S. Department of Agriculture (USDA, 1992).

| Table 4.1 |

POPULATION SIZE AND AVAILABILITY OF RENEWABLE RESOURCES, CIRCA 1990, WITH PROJECTIONS FOR 2010

	1990	2010	Total change	Per capita change
	(million)		(percent)	
Population	5,290	7,030	+33	–
Fish catch (tons)[a]	85	102	+20	−10
Irrigated lands (hectares)	237	277	+17	−12
Cropland (hectares)	1,444	1,516	+ 5	−21
Rangeland and pasture (hectares)	3,402	3,540	+ 4	−22
Forests (hectares)[b]	3,413	3,165	− 7	−30

Sources: Population figures from U.S. Bureau of the Census (1993); 1990 irrigated land, cropland, and rangeland from FAO (1992); fish catch from Perotti (1993); forests from FAO (1992 and 1993) and other sources documented in note 7. For explanation of projections, see text.

[a] Wild catch from fresh and marine waters, excludes aquaculture.

[b] Includes plantations; excludes woodlands and shrublands.

net cropland area will expand much more quickly over the next two decades than it did between 1980 and 1990. Assuming a net expansion of 5 percent, which may be optimistic, the projected 33-percent increase in world population by 2010 would result in a decline of the amount of cropland per person would of 21 percent (see Table 4.1).

PASTURE AND RANGELAND

Pasture and rangeland cover some 3.4 billion hectares of land, more than twice the area in crops. The cattle, sheep, goats, buffalo, and camels that graze them convert grass, which humans cannot digest, into meat and milk, which they can. The global ruminant livestock herd, which numbers about 3.3 billion, thus adds a source of food for people that does not subtract from the grain supply, in contrast to the production of pigs, chickens, and cattle raised in feedlots (FAO, 1993b).

Much of the world's rangeland is already heavily overgrazed and cannot continue to support the livestock herds and management practices that exist today. According to the Global Assessment of Soil Degradation, overgrazing has degraded some 680 million hectares since mid-century. This suggests that 20 percent of the world's pasture and range is losing productivity and will continue to do so unless herd sizes are reduced or more sustainable livestock practices are put in place (Oldeman et al., 1991).

This degradation imposes severe economic costs. In Africa, for example, the annual loss of rangeland productivity is estimated at $7 billion, more than the GNP of Ethiopia and Uganda combined. Asia has by far the largest economic losses from land degradation of any region—an estimated $21 billion per year from waterlogging and salinization of irrigated cropland, erosion of rainfed land, and overgrazing of rangeland (Dregne et al., 1991).

During the eighties, the total range area increased slightly, in part because land deforested or taken out of crops often reverted to some form of grass. If similar trends persist over the next two decades, by 2010 the total area of rangeland and pasture will have increased 4 percent, but it will have dropped 22 percent in per capita terms. In Africa and Asia, which together contain nearly half the world's rangelands and where many traditional cultures depend heavily on livestock, even larger per capita declines will significantly weaken food economies.

Fisheries

Another natural biological system that humans depend on, fisheries add calories, protein, and diversity to human diets. The annual fish catch from all sources, including aquaculture, totalled 97 million tons in 1990, about 5 percent of the protein humans consume. Fish account for a good portion of the calories consumed overall in many coastal regions and island nations.[3]

The world fish catch has climbed rapidly in recent decades, ex-

---

[3] Fish catch from FAO (1993d); percentage of human protein consumption from FAO (1991); contribution to coastal diet from FAO (1993e).

panding nearly fivefold since 1950. But it peaked at just above 100 million tons in 1989. Although catches from both inland fisheries and aquaculture (fish farming) have been rising steadily, they have not offset the decline in the much larger wild marine catch, which fell from a historic peak of 82 million tons in 1989 to 77 million in 1991, a drop of 6 percent.[4]

With the advent of mechanized hauling gear, bigger nets, electronic fish detection aids, and other technologies, almost all marine fisheries have suffered from extensive overexploitation. Under current practices, considerable additional growth in the global fish catch overall looks highly unlikely. Indeed, the U.N. Food and Agriculture Organization (FAO) now estimates that all 17 of the world's major fishing areas have either reached or exceeded their natural limits, and that 9 are in serious decline.[5]

FAO scientists believe that better fisheries management might allow the wild marine catch to increase by some 20 percent. If this could be achieved, and if the freshwater catch increased proportionately, the total wild catch would rise to 102 million tons; by 2010, this would nonetheless represent a 10-percent drop in per capita terms.[6]

FRESH WATER

Fresh water may be even more essential than cropland, rangeland, and fisheries; without water, after all, nothing can live. Signs of water scarcity are now pervasive. Today, 26 countries have insufficient renewable water supplies within their own territories to meet the needs of a moderately developed society at their current population size—and populations are growing fastest in some of the most water-short countries, including many in Africa and the Middle East. Rivers, lakes, and underground aquifers show widespread signs of degradation and depletion, even as human demands rise inexorably (Postel, 1992).

Water constraints are slowing food production, and if present practices and technologies are continued, those restrictions will only

---

[4] Fivefold increase in world fish catch from FAO (various yearsbooks of fishery statistics); other figures from Perotti (1993).
[5] Advent of fisheries technologies and assessment of all fishing areas from FAO (1993e and f).
[6] Estimate of potential catch from an FAO-sponsored book by Gulland (1971); 2010 projection is Worldwatch Institute estimate, based on data from FAO (1993e).

become more severe. Agricultural lands that receive irrigation water play a disproportionate role in meeting the world's food needs: the 237 million hectares of irrigated land account for only 16 percent of total cropland but more than a third of the global harvest. For most of human history, irrigated area expanded faster than population did, which helped food production per person to increase steadily. In 1978, however, per capita irrigated land peaked, and it has fallen nearly 6 percent since then (Postel, 1992).

FORESTS AND WOODLANDS

The last key component of the biological resource base, forests, contribute a host of important commodities to the global economy—logs and lumber for constructing homes and furniture, fiber for making paper, and, in poor countries, fuelwood for heating and cooking. More important even than these benefits, however, are the ecological services forests perform—from conserving soils and moderating water cycles to storing carbon, protecting air quality, and harboring millions of plant and animal species.

Today forests cover 24 percent less area than in 1700—3.4 billion hectares compared with an estimated 4.5 billion about 300 years ago. Most of that area was cleared for crop cultivation, but cattle ranching, timber and fuelwood harvesting, and the growth of cities, suburbs, and highways all claimed a share as well. Recent assessments suggest that the world's forests declined by about 130 million hectares between 1980 and 1990, an area larger than Peru. And as with the other resources, much of the forest resource base is declining in health and quality as well—in part from air pollution and acid rain.[7]

---

[7] The 1700 figure does not include woodlands or shrublands, from Houghton et al. (1983). FAO and the U.N. Economic Commission for Europe/FAO (UNECE/FAO) used somewhat different definitions in their 1990 forest assessments, which precludes a strict comparison of the data between tropical and temperate zones when calculating the change in world forest cover between 1980 and 1990. For both regions, "other wooded areas" have been excluded from all calculations. Because of gaps and discrepancies in the data, FAO does not plan to make any estimate of global deforestation at the conclusion of the 1990 assessment (in progress as of October 1993), according to Janz (1993). The estimated net loss of forests refers to the conversion of forests to an alternative land use, minus the net addition of plantations in tropical regions. As defined by FAO, loss of forests does not include forest that was logged and left to regrow, even if it was clear-cut (unless the forest

REDIRECTING TECHNOLOGY

Like trade, technology has often proven to be a double-edged sword. By allowing for tremendous gains in resource efficiency and productivity, technological advances can help us get more out of each hectare of land, ton of wood, or cubic meter of water.

But for example, the irrigation, agricultural chemicals, and high-yielding crop varieties that made the Green Revolution possible also depleted and contaminated water supplies, poisoned wildlife and people, and encouraged monoculture cropping that reduced agricultural diversity. Huge driftnets boosted fish harvests but contributed to overfishing and the depletion of stocks.

As a society, we have failed to discriminate between technologies that meet our needs in a sustainable way and those that do so by harming the earth. We have let the market largely dictate which technologies move forward, without correcting for its failure to take proper account of environmental damages.

A significant portion of the world's current food output is produced by using land and water unsustainably—in large part because farmers do not pay the full costs of soil erosion, excessive water use, and contamination of the environment by pesticides. For instance, in parts of India's Punjab, the nation's breadbasket, the high-yielding rice paddy–wheat rotation that is common requires heavy doses of agricultural chemicals and substantial amounts of irrigation water. A 1993 study by researchers from the University of Delhi and the World Resources Institute in Washington, D.C., found that in one Punjab district, Ludhiana, groundwater pumping exceeds recharge by one third and water tables are dropping nearly 1 meter per year (Malik and Faeth, 1993).

cover is permanently reduced to less than 10 percent). Thus, the statistics fail to reflect the fragmentation or degradation of forests.

Given these limitations, the 1990 figure is a preliminary and rough Worldwatch Institute estimate, based on FAO (1993g and 1988a, Table 1); all tropical countries from FAO (1993c); Australia, Europe, Japan, and New Zealand from UN-ECE/FAO (1992); Canada from Canadian Council of Forest Ministers (1993) and Lowe (1993); United States from USDA, Forest Service (1982), Waddell, Oswald and Powell (1989), and Bones (1993); former Soviet Union from Shvidenko (1993); China from Smil (1992); Argentina from the National Commission on the Environment (1991); rough estimates for the other temperate developing countries based on trends and projections in FAO (1988b), and in Lanly (1983); area of tropical plantations from FAO (1993c). The estimated loss of 148 million hectares of natural forest was offset by the 18-million-hectare gain in tropical plantations to arrive at the figure of 130-million-hectare net loss between circa 1980 and circa 1990.

Indeed, in many regions—including northern China, southern India (as well as the Punjab), Mexico, the western United States, parts of the Middle East, and much of Africa—water may be much more of a constraint to future food production than land, crop yield potential, or most other factors. Developing and distributing technologies and practices that improve water management is critical to sustaining the food production capability we now have, much less increasing it for the future. Such technologies are available as in the following example:

Water-short Israel is a front-runner in making its agricultural economy more water-efficient. Its current agricultural output could probably not have been achieved without steady advances in water management—including highly efficient drip irrigation, automated systems that apply water only when crops need it, and the setting of water allocations based on predetermined optimum water applications for each crop. The nation's success is notable: between 1951 and 1990, Israeli farmers reduced the amount of water applied to each hectare of cropland by 36 percent (Van Tuijl, 1993).

Paralleling the need for sustainable gains in land and water productivity is the need for improvements in the efficiency of wood use and reductions in wood and paper waste in order to reduce pressures on forests and woodlands. A beneficial timber technology is no longer one that improves logging efficiency (the number of trees cut per hour) but rather one that makes each log harvested go further. Raising the efficiency of forest product manufacturing in the United States, the world's largest wood consumer, to roughly Japanese levels would reduce U.S. timber needs by about a fourth, for instance. Together, available methods of reducing waste, increasing manufacturing efficiency, and recycling more paper could cut U.S. wood consumption in half; a serious effort to produce new wood-saving techniques would reduce it even more (Postel and Ryan, 1991).

THE DOUBLE-EDGE OF TRADE

Imports of biologically based commodities like food and timber are, indirectly, imports of land, water, nutrients, and the other components of ecological capital needed to produce them. Many countries would not be able to support anything like their current population and consumption levels were it not for trade.

In principle, there is nothing inherently unsustainable about one nation relying on another's ecological surplus. The problem, however, is the widespread perception that all countries can exceed their carrying capacities and grow economically by expanding manufactured and industrial goods at the expense of natural capital—paving over agricultural land to build factories, for example, or clear-cutting forest to build new homes. But all countries cannot continue to do this indefinitely. Globally the ecological books must balance.

Many economists see no cause for worry, believing that the market will take care of any needed adjustments. As cropland, forests, and water grow scarce, all that is necessary, they say, is for prices to rise; the added incentives to conserve, use resources more productively, alter consumption patterns, and develop new technologies will keep output rising with demand. But once paved over for a highway or housing complex, or eroded, cropland is unlikely to be brought back into production—no matter how severe food shortages may become. Moreover, no market or price mechanism exists for assuring that an adequate resource base is maintained to meet needs that the marketplace ignores or heavily discounts, including those of vital ecosystems, other species, the poor, or the next generation.

Trade in forest products illuminates some of these trends. East Asia, where the much-touted economic miracles of Japan and the newly industrializing countries have taken place, has steadily and rapidly appropriated increasing amounts of other nations' forest resources. In Japan, where economic activity boomed after World War II, net imports of forest products (expressed in equivalent units of wood fiber content) rose eightfold between 1961 and 1991 (see Table 4.2). The nation is now the world's largest net importer of forest products by far. Starting from a smaller base, South Korea's net imports have more than quadrupled since 1971, and Taiwan's have risen more than sevenfold.[8]

China is a big wild card in the global forest picture. According to He Bochuan, lecturer at Sun Yat-sen University in Guangdong, China's consumption of raw wood—some 300 million cubic meters per year—exceeds the sustainable yield of its forests and woodlands by 30 percent. During the last decade alone its net forest product imports more than doubled. With one-fifth of the world's population, economic growth rates that have averaged around 12 percent in recent

---

[8] Worldwatch Institute estimates, based on FAO (1993a); Taiwan from Wardle (1993b); broad conversion factors for converting non-roundwood products into green solid wood equivalent—*not roundwood raw material equivalent*—were supplied by Wardle (1993a).

Table 4.2

NET IMPORTS OF FOREST PRODUCTS, SELECTED EAST ASIAN COUNTRIES, 1961-91

Country	1961	1971	1981	1991
	(thousand cubic meters)[a]			
Japan	8,800	45,000	50,000	70,100
South Korea	500	2,900	4,700	12,700
Taiwan	0	1,200	6,700	8,800
China	200	200	3,100	6,800
Hong Kong	900	1,300	2,000	2,500
Singapore	200	1,600	1,400	1,100

Sources: Worldwatch Institute based on FAO (1993a); Taiwan from Wardle (1993b); broad conversion factors for converting non-roundwood products into green solid wood equivalent, which differs from roundwood raw material equivalent, were supplied by Wardle (1993a).
[a] All forest products are expressed in equivalent units of wood fiber content.

years, and its own limited stocks undergoing depletion, China is now becoming the leading importer of wood. If its per capita use of forest products were to rise to the level of Japan's today, China's total demand will exceed Japan's by nine times—and its import needs will place enormous pressure on forests worldwide.[9] Thus, trade cuts both ways. It can help overcome local or regional carrying capacity constraints by allowing countries to bring in resources to meet their needs. But it can also foster unsustainable consumption levels—and environmental destruction—by creating the illusion of infinite supplies.

LIGHTENING THE LOAD

Ship captains pay careful attention to a marking on their vessels called the Plimsoll line. If the water level rises above the Plimsoll line, the

[9] Sustainable yield of forests and woodlands from He Bochuan (1991); rise in Chinese demand is Worldwatch Institute estimate for 1991, based on sources and methodology cited in note 9.

boat is too heavy and is in danger of sinking. When that happens, rearranging items on the ship will not help much. The problem is the total weight, which has surpassed the carrying capacity of the ship.[11]

Herman Daly sometimes uses this analogy to underscore that the scale of human activity can reach a level that the earth's natural systems can no longer support. The ecological equivalent of the Plimsoll line may be the maximum share of the earth's biological resource base that humans can appropriate before a rapid and cascading deterioration in the planet's life-support systems is set in motion. Given the degree of resource destruction already evident, we may be close to this critical mark. The challenge, then, is to lighten our burden on the planet before, metaphorically speaking, the ship sinks.

The days of the frontier economy, in which abundant resources were available to propel production growth and living standards, are clearly over. We have entered an era in which global prosperity increasingly depends on using resources more efficiently, distributing them more equitably, and reducing consumption levels overall. Unless we accelerate this transition, we risk exceeding the planet's carrying capacity to such a degree that a future of economic and social decline will be impossible to avoid.

[10] This analogy is borrowed from Herman Daly, senior economist, World Bank. See Daly (1992).

CHAPTER 5

Failures of the System of National Accounts

National accounting, Wilfred Beckerman has written in a standard introduction to the topic, is simply a systematic way of classifying the multitude of economic activities that take place in the economy in different groups or classes that are regarded as being important for understanding the way the economy works (Beckerman, 1968, p. 68). Beckerman acknowledges that there is an arbitrary element in many of the decisions that have to be made in drawing up a classification system for the National Accounts, and he is at pains to stress that there is nothing sacrosanct about the present way the accounts are structured:

It must be emphasised that there is a constant evolution and change in the questions the economists are asking, in the institutional structure of the economy, and in the working hypotheses that economists use for purposes of analysing the behaviour of the economy. In accordance with these changes, so it will be necessary to modify and adapt the classification system used for national accounting. It would be useless to persist with a classification system, for example, that no longer corresponded to the institutional and social categories of society, or to the latest knowledge about how the economy operated and so about which relationships were important for analytical purposes. . . . It is to be expected that the appropriate National Accounts classification will, as the years go by, be subject to far-reaching modifications [ibid., pp. 5–6].

This is also the view of Robert Eisner, who noted in his authoritative survey of proposals to amend or extend the National Accounts: "(T)he accounts are not set in concrete. . . . From the early days of their formulation there have been lively debates as to just what they should include, how items ought to be measured, and how they are to be put together. They have been modified over the years" (Eisner, 1988, pp. 1611–12).

These quotations from definitive texts on national accounting show that proposing to adjust the National Accounts for newly important economic facts or perceptions is neither subversive nor eccentric. Rather it is indicative of continuing open-mindedness in economic science, whereby even the most settled procedures are open to modification if new conditions render it desirable. The new conditions in this case are the myriad ways in which economic activity is degrading the environment, as we have stated before.

When the National Accounts were being systematised in the 1940s, environmental issues had a low perceived importance, and the accounting structure adopted simply ignored environmental issues, as one of the essentially arbitrary judgements referred to by Beckerman. The National Accounts have now matured into the fundamental instrument of macroeconomic management and the means of indicating economic progress. It is frankly inconceivable that national strategies for sustainable development will cut any ice economically unless they are integrally related to the national accounting system. The European Community certainly recognised the importance of this when, in its Fifth Action Program on the Environment, it stated categorically that environmentally adjusted National Accounts should be available on a pilot basis from 1995 onwards for all Community countries, with a view to formal adoption by the end of the decade (EC, 1992, p. 97).

This political imperative for environmental adjustment of the National Accounts reinforces the intellectual and economic imperatives. The National Accounts' current treatment of environmental issues can only be justified intellectually if these issues are economically unimportant. If these issues are now crucially important to humanity, and are intimately related to economic activity, as the Rio Summit made clear, then the National Accounts' treatment of them is bizarre and indefensible. Moreover, it is economically misleading. As discussed below, a number of studies (e.g. Repetto et al., 1989 for Indonesia, Solórzano et al., 1991 for Costa Rica, Adger, 1992 for Zimbabwe, Van Tongeren et al., 1993 for Mexico, Bartelmus, Lutz and Schweinfest, 1993 for Papua New Guinea) have shown that GNP growth rates and net capital formation can be greatly overstated by failing to account properly for natural resources.

In short, for intellectual, economic and political reasons, the adjustment of the National Accounts for environmental factors has now become a priority.

THE IMPORTANCE OF THE ENVIRONMENT TO PRODUCTION

The first, and perhaps most important, question the National Accounts are intended to answer is: "How much does the national economy produce?" To answer this question the National Accounts must be, and are, based on a clear theory of production and definition of national product, which, according to Beckerman, is "the unduplicated value of the flow of goods and services produced by the nation in the time period concerned (usually a year)" (Beckerman, 1968, p. 31). If the environment is to be included in the accounts, then it must be incorporated consistently into this theory and definition.

Current theory[1] abstracts from the physical reality of production.

---

[1] The economic theory of production uses the concept of a production function, whereby output is produced from a combination of inputs. A common production function used in growth theory (e.g. Solow, 1991) is $Y = F(K,L,t)$, where Y is output, K is capital inputs, L is labor inputs, t is time (introduced to reflect technical progress) and F(.) is the function which states how the inputs are combined.

It even omits the third commonly recognised factor of production, *land* (capital and labor being the other two), so that there is no scope for analysing the environmental contribution to production in such a model. As Nordhaus and Tobin have said: "[The prevailing standard model of growth] is basically a two factor model in which production depends only on labor and reproducible capital. Land and resources, the third member of the classical triad, have generally been dropped. The simplifications of theory carry over into empirical work. The thousands of aggregate production functions estimated by econometricians in the last decade are labor-capital functions" (Nordhaus and Tobin, 1973, p. 522).

More recently there have been attempts to incorporate at least some aspects of the environmental contribution. Thus the work of Jorgenson and Wilcoxen (1993) implies a production function, where the new variables are energy and intermediate material inputs respectively.[2] Such a formulation obviously enables the contribution of energy to production to be studied but it still omits an analysis of the environment's wider contribution.

Figure 5.1 is a diagrammatic representation of a production process which includes "environmental flows" rather than just those of energy. The diagram is adapted from Ekins (1992) and is consistent with the description of production by Harrison (1993, pp.25ff.) and Bartelmus and Tardos (1993, p. 185), in connection with their discussion of integrated economic and environmental accounting.

The figure portrays three kinds of capital stock: ecological (or natural) capital, human (individual and social) capital, and manufactured capital. Each of these stocks produces a flow of "services"—environmental, E, labor, L, and capital, K, services—which serve as inputs into the productive process, along with "intermediate inputs", M, which are previous outputs from the economy which are used as inputs in a subsequent process.

Manufactured capital comprises material goods—tools, machines, buildings, infrastructure—which contribute to the production process but do not become embodied in the output and, usually, are "consumed" in a period of time longer than a year. Intermediate goods, in contrast, either are embodied in produced goods (e.g. metals, plastics, components) or are immediately consumed in the production process (e.g. fuels). Human capital comprises all individuals'

[2] $Y = F(K,L,E,M,t)$

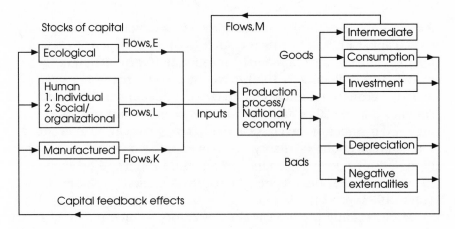

Figure 5.1 Stocks and flows in the process of production

capacities for work (skills, knowledge, health, strength, motivation) and the networks and organizations through which they are mobilised.

Ecological capital is a complex category which performs three distinct types of environmental function, two of which are directly relevant to the production process. The first is the provision of resources for production, the raw materials that become food, fuels, metals, timber etc. The second is the absorption of wastes from production, both from the production process and from the disposal of consumption goods. Where these wastes add to or improve the stock of ecological capital (e.g. through recycling or fertilisation of soil by livestock), they can be regarded as investment in such capital. More frequently, where they destroy, pollute, or erode, with consequent negative impacts on the ecological, human or manufactured capital stocks, they can be regarded as agents of negative investment, depreciation, or capital consumption.

The third type of environmental function is not shown here because it does not contribute directly to production, but in many ways it is the most important type because it provides the basic context and conditions within which production is possible at all. It comprises basic survival services such as those producing climate and ecosystem stability, shielding of ultraviolet radiation by the ozone layer, and amenity services such as the beauty of wilderness and other natural areas. These services are produced directly by ecological capital independent of human activity, but human activity can certainly have an (often negative) effect on the responsible capital and therefore on the services produced by it.

The outputs of the economic process can, in the first instance, be categorised as Goods and Bads. The Goods are the desired outputs of the process, as well as any positive externalities (incidental effects) that may be associated with it. These Goods can be divided in turn into consumption, investment, and intermediate goods and services. The Bads are the negative effects of the production process, including capital depreciation and negative externalities, such as those contributing to environmental destruction, negative effects on human health, etc. Insofar as they have an effect on the capital stocks, the Bads can be regarded as negative investment.

The necessity for a matter/energy balance on either side of the production process means that all matter and energy that feature as inputs must also emerge as outputs, either embodied in the Goods or among the Bads. On disposal of the former, therefore, all these former inputs are returned to the environment, to the stock of ecological capital, where they may have a positive, negative or neutral effect. The essential point is that, for matter, Figure 5.1 represents a closed system; for energy, inputs can be received from the sun, and heat can be radiated from the earth into space.

CONSTRUCTION OF THE NATIONAL ACCOUNTS

Figure 5.1 identifies as a production process any humanly-organized combination of flows from the capital stocks which results in the production of desired outputs, Goods. It makes no distinction between flows of inputs that are paid for and those that are not; or between outputs that are marketed and those that are not. However, these distinctions become important when one wishes to quantify the flows and outputs, which is, of course, the principal purpose of the National Accounts, to arrive at a figure for national product.

The National Accounts are presently constructed by defining a production boundary, which distinguishes between productive and non-productive activities. As Beckerman notes, "the dividing line between what is and what is not a productive activity is necessarily arbitrary in any system of National Accounts" (op. cit., p. 8). The current dominant convention is that where the flows of goods and services are accompanied by corresponding flows of money (in the opposite direction), then they are part of the national product. Where they are not thus accompanied, they are not part of the national product. The principal exceptions to this rule are the residential services provided by buildings to owner-occupiers and the home-grown food consumed by farmers, the values of both of which are estimated (imputed) according to reigning market prices and added into the national product. With few such exceptions, a flow of goods and services that is not accompanied by a money flow is not considered part of the national product, or GNP. Similarly, any flow of money that is not matched by a flow of goods and services (such as the payment by government of old-age pensions or unemployment benefits) is called a transfer payment and is not added to the national product.

Finally, care must be taken in the accounting of intermediate goods and services. If their value is counted as part of national product; and if the value of the goods and services to the production of which they contribute (which will include their value) is also counted as part of the national product, then the value of the intermediate goods will have been counted more than once—double-counting will have occurred. Therefore the national product must be counted either from the value only of final goods and services, those which are sold directly for consumption and investment (which will include the value of all intermediate goods); or it must be calculated by summing the value-added from each industry, the payments only for capital (including rents) and labor services in each industry. National product computed in either of these ways will include the value of intermediate goods only once.

This method of constructing national product is shown in Figure 5.2. The gray lines show the flow of goods and services into and out of the production sector, which produces intermediate, consumption, investment, and export goods. The other lines show flows of money, in the reverse direction to the flow of goods. Households comprise people as workers, who provide labour services (L), and as owners (of firms and property), who provide capital services (K,E', where E' is that part of ecological capital which is marketed). For these services, workers are paid wages (W), shareholders receive profits (P), and property-owners receive rents (R). These comprise household factor incomes. There then take place various transfer payments between people and government (taxes, T; and government benefits, Tr), from which households and governments emerge with net incomes, (H and G respectively). Some of this income is spent on imports (M). Otherwise households purchase domestically produced consumption (C)

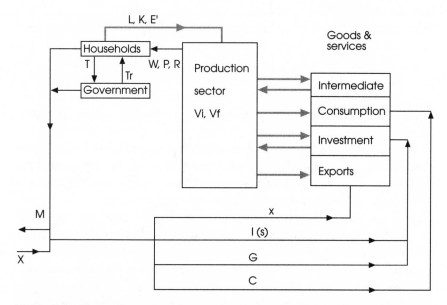

Figure 5.2 Flows of money as they are entered into the National Accounts

or investment (I) goods, according to whether they spend or save (S) their incomes. Government expenditure (G) is also spent on investment or consumption goods. Exports are bought with the money of foreigners (X).

The flows of money, adjusted for imports and exports, are circular: the production sector pays out incomes from the value-added which it generates, which it receives from expenditures on its products. The value of the national product is the value of the expenditures on it, but, because of the circular flow of the money concerned, this must be equal to the factor incomes and the value-added generated in production. The figure obtained by calculating any of these totals is called Gross Domestic Product (GDP). GNP is obtained from GDP by adding in net income from abroad.

The use of the term "income" as above, while usual, is not strictly correct. The strict definition is that of John Hicks: "We ought to define a man's income as the maximum value which he can consume during a week, and still expect to be as well off at the end of the week as he was at the beginning" (Hicks 1946, p. 172). The factor incomes identified above are not income in this sense, because no allowance has been made for the capital depreciation that has taken place in generating them. The National Accounts recognise this and calculate a figure for this depreciation, or capital consumption, deducting it from

GNP to arrive at Net National Product (NNP), which is also called National Income.

CAUSES FOR CONCERN IN THE NATIONAL ACCOUNTS

Comparing the full model of production described here and the method of calculating the National Income, it can be seen that there are three major ways in which the National Accounts fail to give a correct assessment of income, as defined by Hicks.

The first is the summary exclusion of practically all nonmonetary production, so that GNP as calculated will be substantially less than a Total Product which included nonmonetary (including household) production. But the error introduced by excluding nonmonetary production goes further than this, for monetary and nonmonetary production are not independent of each other. One of the main characteristics of industrial economic development has been the transfer of activities (e.g. cooking, washing, cleaning, child-care, education) from the household, family, and community, where they were not paid for, to market-settings (e.g. restaurants, launderettes, cleaning firms, nurseries, schools) where they are paid for. The National Accounts calculated as above show the whole of the new market activity as growth in production, whereas the actual net product of the transfer is the difference between the value of the new market activity and the value of the nonmonetary production it replaced.

It was to avoid an anomaly of this sort that the decision was taken to include in national product the imputed rents of owner-occupied households. Otherwise GDP would have decreased by the amount of the rent every time a tenant bought the house he or she was renting (and increased every time an owner-occupier became a tenant). This would obviously have been an absurd situation. It is, however, precisely analogous to the situations identified by Pigou (1932, pp. 32–33), such as that whereby, under current conventions, GDP decreases by the amount of her pay when a man marries his housekeeper, even if she performs the same tasks after her marriage as before, though without a wage. More contemporaneously, one might ask by how much GDP overstates the increase in production when a married couple is divorced, and both individuals get full-time jobs and start paying for services which before they rendered freely to each other.

Therefore the exclusion of nonmonetary production from GDP

not only understates real production. Given the shift of activities towards the market that occurs with modern economic development, it also systematically overstates economic growth. It also denies the value of nonmonetary work, which discourages and discriminates against those who do it, still mainly women. For all these reasons nonmonetary production should be brought into the National Accounts. Goldschmidt-Clermont (1992, 1993) has described clearly how this could be achieved by shifting the production boundary to include these activities as productive, and then calculating their value, on the basis of comparison with their nearest market equivalents, in accordance with current national accounting conventions. The implementation of such a method is overdue, but is, perhaps, in view. As Goldschmidt-Clermont (1993, p. 419) notes: "According to the currently available information, the 1993 SNA recommends the inclusion of part of households' non-market production within the production boundary." (The other part is recommended to be included in a satellite account, as discussed below.)

The second major area of omission in the National Accounts is their failure to assess changes in human (including social and organizational) capital. Growth accounting has now developed very detailed ways of differentiating between different qualities of labour input. For example, the model described in Jørgenson (1990) cross-classifies labor inputs by the two sexes, eight age groups, five educational groups, and two employment statuses—employee and self-employed (Jørgenson, 1990, p. 35). Clearly these different qualities of labour input, which have different productivities, issue from a heterogeneous human capital stock, which could, in principle, be valued either according to the productivity of associated labor services, as expressed by the present value of the streams of incomes to which the stock gives rise; or according to the inputs required to create, maintain, and increase it. In practice, however, the task appears too ambitious. While it is beyond contention, for example, that malnutrition degrades human capital, that does not mean that all food expenditures should be regarded as investment in such capital, nor would it be possible to distinguish between food consumption and food investment without making a plethora of arbitrary assumptions and running into huge data problems. Expenditures on health and education, the other principal components of human capital creation and maintenance, present a similar problem.

If human capital is statistically intractable, social and organizational capital is even more so. Harrison is in no doubt about the importance of this kind of capital, which she calls institutional capital: "Economic behavior does not take place in a vacuum. Rather it consists of a large number of agents interacting according to a known set of norms and conventions. These are enshrined in the laws and customs of the country through the formal and informal activities of the governing body" (Harrison, 1993, p. 26). In addition to political and legal institutions, social and organizational capital includes firms, business associations, trade unions, voluntary groups, and families; for all these organizations play an important role in monetary and nonmonetary production. But, as Harrison says: "Just because the importance of institutional capital is recognized, however, does not mean it can be represented in economic terms. . . . Even less than human capital (to which of course it is linked) can it be associated with specific expenditures or approximated in money terms" (ibid., p. 26).

The issues of human capital and non-market production are discussed in great detail in Eisner's 1988 survey of proposed national accounting extensions. What is notable is that Eisner practically ignored the third great area of failure of the National Accounts, their treatment of the environment. Partly this was due to the lack of worked-out environmental adjustments for Eisner to include in his empirical review. But the virtual omission of the environment from the conceptual and methodological discussion also indicates the low profile this issue had among national income accountants at that time. The change of attitude in relatively few years has been striking, and restores the issue of the treatment of the environment in the National Accounts to an importance it had in an earlier period of environmental awareness, the 1970s. Then, one of the fathers of national accounting had written:

> Pollution, depletion and other negative by-products of economic production were greatly accelerated because of the very rapid rise of total output. . . . Economic production, and the technology that it employs, may be viewed as interference with the natural course of events, in order to shape the outcome to provide economic goods to man. All such interference has potential negative economic consequences—pollution and the like—the more lasting, the higher the level of production technology as measured by its capacity to produce goods [Kuznets, 1973, p. 585].

Herfindahl and Kneese (1973, pp. 447–48) were confident about the implications of these "negative ecological consequences" for the National Accounts:

The exclusion of the services of clean air, clean water, space etc. from the list of final goods probably is not the result of disagreement that the services provided by nature are a factor in true welfare, but rather of the judgment on the part of the income accountant that obtaining acceptable estimates for these values would be too difficult and costly. It is clear, however, that any reduction in the service flows of common property resources that is viewed as a loss of real product by consumers means that NNP overstates any increase in final product as compared with the total flow from the truly relevant and larger list of final goods and services. In the extreme case the 'true' service flow could actually decrease while NNP rises.

Juster (1973, p. 66) agrees: "The environment is clearly worse today than it was in the mid-1950s, and comparison of real output between these two periods is already over-stated because environmental deterioration has been permitted to occur."

This effect can be considered in terms of the concept of ecological capital developed earlier. As was then seen, the economic functions of the environment include the provision of resources, the absorption and neutralization of wastes, and the provision of other services independent of human agency. In the process of economic activity, resources can become depleted (and, in the case of non-renewable resources, are bound to become so) and the environment can become degraded through pollution or occupation (change of use). Exactly analogous to the consumption of manufactured capital, continuing depletion and degradation imply that the environment will in the future be able to fulfill its functions less effectively—or not at all. Yet in many cases these functions are of vital importance to economic, and wider human, life. It is for this reason that current use of the environment is perceived as unsustainable: it cannot be envisaged to continue.

The calculation of GNP and National Income (NNP) give no inkling of this increasingly serious unsustainability, which is far more pronounced than in the early 1970s, despite the fact that, as discussed earlier, the standard definition of income is as a sustainable quantity. It is this inconsistency in the accounts' treatment of the environment that is one of the components of the recent upsurge of concern over the issue, which has led one economist, Jan Tinbergen, a Nobel Laureate who, like Kuznets, was among those responsible for the early development of the accounts, to say that, because of them, society is steering by the wrong compass (Tinbergen and Hueting, 1993, p. 52).

One of the differences between the 1990s and the 1970s is that there now exist a number of quantitative estimates of adjustments that would need to be made to the National Accounts for them to reflect environmental depletion and degradation. The detail of how the adjustments have been calculated will be discussed in a later chapter. Here we'll report some of the estimates themselves to give an idea of the quantities involved.

The aim of estimating these numbers is to make some assessment of the value of the ecological capital stock that has been depleted or degraded as a result of the processes of production and consumption, i.e. as a result of economic activity. This restriction of the focus to economic activity is simply because this is the focus of the National Accounts. Any attempt to adjust the accounts must start from the same area of concern.

Bartelmus et al. (1993, p. 108) set out the three basic shortcomings of the National Accounts with regard to the environment: National Accounts have certain drawbacks that cast doubt on their usefulness for measuring long-term environmentally sound and sustainable economic development. First, they neglect the scarcities of natural resources that can pose a serious threat to sustained economic productivity. Second, they pay only limited attention to the effects of environmental quality on human health and welfare. And, third, they treat environmental protection expenditures as increases in national product, which could instead be considered social costs of maintaining environmental quality.

With regard to the first of these points, the depletion of natural resources, several studies have now computed a partial estimate of these. Van Tongeren et al. (1993) calculated the value of the depletion of oil and forests in Mexico in 1985. They found that this depletion amounted to 5.7 percent of Mexico's GDP. More serious, perhaps, for Mexico's future production prospects, they found that net capital accumulation fell from 11 percent to less than 6 percent of NDP when depletion was accounted for. The sectoral effects are, predictably, even more pronounced. Forestry and oil extraction made a contribution to unadjusted NDP of 0.54 and 3.50 percent respectively. When depletion was taken into account their contribution fell to 0.15 and 0.00 percent—that is, this study accounted all of oil extraction's contribution to NDP as capital consumption rather than the generation of income (figures from Van Tongeren et al., 1993, Table 6.9, p. 100; Table 6.12, p. 105).

Adger (1992) produced estimates for Zimbabwe for 1987. He

found that depletion of ecological capital due to soil erosion and forest loss reduced the net product of Zimbabwe's commercial and communal agriculture sector by 30 percent and Zimbabwe's NDP by 3 percent. Adger further analyzed the true economic results of a surge in minerals production in 1991 following the devaluation of the Zimbabwean dollar. In nominal terms, the sector's conventionally calculated net product grew by 18 percent; but depletion grew from 20 to 27 percent of this net product. When this is taken into account, the adjusted net sectoral product only grew by 7 percent (rather than 18 percent). In real terms, the sectoral growth was 5 percent rather than the conventionally calculated 7 percent (figures from Adger, 1992, Table 4, p. 19).

The World Resources Institute has also done two studies of this kind. Solórzano et al. (1991) found for Costa Rica that the loss of three resources—forests, soil and a fishery—grew from 5.67 percent of NDP in 1970 to 10.5 percent in 1988. Capital formation was even more drastically affected: net capital formation was lower by 38 percent in 1970 and 48 percent in 1988, if depletion is taken into account (figures from Solórzano et al., 1991, Table 1.2, p. 7, Table 1.3, p. 9). For Indonesia, Repetto et al. (1989, p.6) found that subtracting the depletion of oil, forests and soil from the country's GDP reduced its average growth rate over the years 1971-1984 from 7.1 to 4.0 percent. The study also showed that in some years, using these calculations of the depletion of just these three resources and contrary to the conventional accounts, net investment in the Indonesian economy was negative: A fuller accounting of natural resource depletion might conclude that in many years depletion exceeded gross investment, implying that natural resources were being depleted to finance current consumption expenditures (ibid., p. 6).

Repetto et al. (ibid., pp. 2-3) spell out the implications of their study, which are also applicable to the other studies cited;

Under the current system of national accounting, a country could exhaust its mineral resources, cut down its forests, erode its soils, pollute its aquifers, and hunt its wildlife and fisheries to extinction, but measured income would not be affected as these assets disappeared. . . . (The) difference in the treatment of natural resources and other tangible assets confuses the depletion of valuable assets with the generation of income. . . . The result can be illusory gains in income and permanent losses in wealth.

Bartelmus et al.'s second basic environmental shortcoming of the National Accounts is their failure to account properly for economically-induced environmental degradation and its effects on human health and welfare. The effects on human health and welfare can entail two types of costs: actual monetary expenditures incurred to defend against the effects; and losses of health, production, or welfare due to a lack of such defensive expenditures. The defensive expenditures due to degradation are simple to calculate in principle because they represent actual monetary flows, although data in the required disaggregation may not always be available. The other losses are clearly a more difficult category of ecological capital consumption to measure than either the defensive expenditures or the depletion of such resources as oil, timber, and fish that was investigated earlier. Because there are no corresponding monetary flows, monetary values have to be imputed to these losses using a variety of methods, none of which is totally satisfactory.

The study by Van Tongeren et al. (1993) that has already been cited makes an estimate of the costs for Mexico in 1985 of a variety of forms of land, water, and air degradation and pollution. The authors find that these costs are 7.6 percent of NDP and no less than 67 percent of net capital accumulation. When these costs are added to the resource depletion costs given earlier, to give a figure of ecological capital consumption that includes both depletion and degradation, this figure is 13.4 percent of NDP and capital accumulation becomes − 2.3 percent of NDP; i.e., instead of the economy having net investment of 11.2 percent, as the conventional accounts showed, it had net disinvestment of 2.3 percent, when ecological capital was accounted for. This is a chronic situation of economic, as well as environmental, unsustainability.

Another calculation of the costs of ecological degradation comes from Cobb and Cobb (1994) for the United States, as part of their calculation of an Index of Sustainable Economic Welfare (ISEW), which also included the calculation of some depletion costs. For 1986 their calculated costs of air, water and noise pollution plus long-term environmental degradation amounted to 17 percent of U.S. GNP (15 of the 17 percent was long-term environmental degradation, which was a largely speculative figure, but the pollution costs were conservative). Their costs of depletion of non-renewable resources and loss of farmland and wetlands amounted to 7 percent of U.S. GDP. Thus, even leaving long-term degradation out of account completely, ecological capital equivalent to about 10 percent of U.S. GDP was consumed in 1986 (figures from Cobb and Cobb, 1994, Table A.1, pp. 82–83).

Interestingly, Cobb and Cobb find that their ISEW decreased by 7 percent from 1980 to 1986, although GNP rose by 11.6 percent over the same period, giving an empirical example of the theoretical possibility noted by Herfindahl and Kneese, as quoted above, 20 years ago. ISEW is further discussed, in the version of Daly and Cobb (1989), which comes to the same general conclusions, in Chapter 10.

Bartelmus et al.'s third category of National Accounts anomalies was the treatment of environmental protection expenditures. These are expenditures that prevent environmental degradation rather than trying to defend against or compensate for it. However, they are normally included in calculations of defensive expenditures. Leipert (1989a, Table 1, p. 852) has calculated that environmental protection expenditures in Germany in 1985 were 1.48 percent of GDP. The costs of environmental degradation, excluding health costs, were less, 0.80 percent of GDP (See Chapter 11). This is in stark contrast to the Mexican case, where Van Tongeren et al. (1993, p. 98) estimated that environmental protection services were only 5 percent of the value of the environmental degradation cost.

There is a long and still unresolved debate over the correct treatment of defensive expenditures in the National Accounts. But, however they are to be accounted, no one denies that they are costs arising from unintended effects of production and consumption. If the economic activity was not going to or did not bring about the unintended effect, then there would be no need to incur the costs and society would be better off.

Finally, it must be remembered that in all the studies cited above the authors are cautious about the accuracy of the actual numbers calculated. However, it is also true that they only try to measure a small fraction of the environmental depletion and degradation, in the economies concerned, so that the total ecological capital consumption could be very much higher than their estimates. To arrive at the full environmental costs incurred by the economy, omitted health and welfare effects of the depletion and degradation would have to be considered as well.

From PRODUCTION TO WELFARE

The view of two authors from the UK's Central Statistical Office is blunt: The National Accounts measure activity involving economic exchanges. They do not measure, nor claim to measure, sustainable development or welfare (Bryant and Cook, 1992, p. 99). It is worth quoting at some length Herfindahl and Kneese's exposition of the conditions for NNP to be a useful welfare indicator.

> If it were true that all salient goods and services were exchanged in markets; that the degree of competition in these markets did not change; that the programs of government and non-profit institutions did not change in such a way that they produced substantially altered welfare relative to the final goods and services that they absorbed; that population stayed constant; and that the distribution of income did not change; then alterations in real NNP . . . could be regarded as a good indicator of changes in the economic welfare of the population.
>
> This is an imposing string of assumptions, none of which is ever exactly met in reality. . . . In fact, the distance between reality and these assumptions is large and significant in some cases, thus seriously reducing the practical usefulness of NNP as a welfare measure. . . . The designers of the accounts thought that at best they would serve to provide only a rough indicator of one dimension of welfare [Herfindahl and Kneese, 1973, p. 446].

Despite the theoretical consensus on this matter, GNP and NNP continue to be widely used as indicators of welfare, both in popular discourse and in academic work. Beckerman provides a clear example of academic ambivalence on this issue. He is clear that income and welfare are not the same thing: "With exactly the same income and prices and quantities today as yesterday, and without any change in my tastes. . . . I may still be less happy today than yesterday on account of some change in economic circumstances" (Beckerman, 1968, p. 167). Yet five pages later, without so much as an invocation of *ceteris paribus*, he is proposing to discuss how to measure differences in real income, *"which we will equate with economic welfare"* (ibid., p. 172, emphasis added).

Examples of the use of GDP as a welfare measure abound in practical work. Thus, to take the global warming issue, Boero et al. report in their survey of the CO_2 abatement literature: "Even if consumption were proportional to GDP, welfare is not proportional to consumption, and so measuring the costs of abatement in terms of GDP is not ideal. However, given that this is *the almost universal metric*, GDP is the main focus of this survey" (Boero et al., 1991, p. S3, emphasis added).

This issue was addressed explicitly by Nordhaus and Tobin in a seminal paper in 1973:

> GNP is not a measure of economic welfare. . . . An obvious short-coming of GNP is that it is an index of production, not consumption. The goal of economic activity, after all, is consumption. Although this is the central premise of economics, the profession has been slow to develop, either conceptually or statistically, a measure of economic performance oriented to consumption, broadly defined and carefully calculated (Nordhaus and Tobin, 1973, p. 512).

Yet, in his influential paper on the economics of the greenhouse effect 18 years later, Nordhaus explicitly employs the GDP metric as a welfare measure: We assume that it is desirable to maximise a social welfare function that is the discounted sum of the utilities of per capita consumption (Nordhaus, 1991, p. 925). In the paper consumption is expressed in terms of GDP.

In order to relate welfare to the earlier discussion of production, Figure 5.3 introduces welfare into the model of Figure 5.1. Welfare is seen to be affected by many factors apart from the consumption of produced goods and services, including the process of production (working conditions), the institutions of production (family, commu-

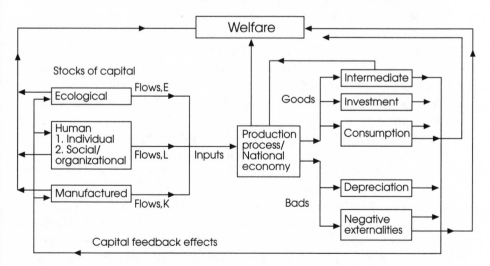

Figure 5.3 Stocks, flows, and welfare in the process of production

nity, political/legal system), people's state of health (part of human capital), and environmental quality (negative externalities and the stock of ecological capital).

It is of doubtful usefulness to seek to define consumption to include the effects of all these factors, as Nordhaus and Tobin recommend (see above). It seems more satisfactory to acknowledge that economic welfare is a function of the stocks, flows and processes related to production, and of its social, economic, and environmental outcomes (including consumption and income distribution). This seems to be the approach taken by Pearce et al., who consider development—implying change that is *desirable* (and therefore equivalent to an increase in welfare)—to be a *vector* of desirable social objectives; that is, it is a list of attributes which society seeks to achieve or maximize. The elements of this vector might include the following:

- increases in real income per capita;
- improvements in health and nutritional status;
- educational achievement;
- access to resources;
- a fairer distribution of income; and
- increases in basic freedoms.

Correlation between these elements, or an agreed system of weights to be applied to them, might permit development to be represented by a single proxy indicator, but this is not an issue pursued here (Pearce et al., 1990, pp. 2-3). A very similar approach was taken by Hueting (1986, pp. 243ff.), who identified welfare as having the following components: production (income), environment, employment, working conditions, income distribution, leisure and safety of the future.

Much of the 1973 paper by Nordhaus and Tobin in fact comprised an attempt to calculate a proxy welfare measure, called Measure of Economic Welfare (MEW), through some reclassification of GNP final expenditure, imputations for capital services, leisure and non-market work, and deductions for disamenities of urbanization. The detail of the calculation need not concern us here, but Nordhaus and Tobin's conclusion is significant, in that it seemed to confirm a rough correlation between GDP and welfare: Although GNP and other national income aggregates are imperfect measures of welfare, the broad picture of secular progress which they convey remains after correction of their most obvious deficiencies (ibid., p. 532).

Seventeen years later Daly and Cobb came to a very different conclusion, both by interpreting Nordhaus and Tobin's figures differently, and on the basis of their own calculation of an Index of Sustainable Economic Welfare (ISEW). With regard to the former they note: "When (Nordhaus and Tobin's) findings are more carefully examined for time frames other than the full period from 1929–65, the relatively close association between per capita GNP and MEW disappears" (Daly and Cobb, 1990, p. 79). In particular, if the period 1947–65 is chosen, the difference between the growth of GNP and MEW (both per capita, hereafter PC-GNP and PC-MEW) is striking: 48 or 2.2 percent per annum for PC-GNP, 7.5 or 0.4 percent per annum for PC-MEW. This leads Daly and Cobb to conclude: "With their own figures, Nordhaus and Tobin have shed doubt on the thesis that national income accounts serve as a good proxy measure of economic welfare" (ibid., p. 80).

ISEW makes different adjustments to GNP to MEW, including giving consideration to resource depletion and environmental damage, so the two indices are not strictly comparable. However, the general trend in ISEW is similar to that of MEW for the same years: PC-GNP growth from 1950–86 was 2.02 percent, while PC-ISEW grew by 0.87 percent. From 1970–86, while PC-GNP maintained roughly its 2 percent growth rate, PC-ISEW actually declined, at an increasing rate (-0.14 percent from 1970–80; -1.26 percent from 1980–86) (ibid., p. 453).

From an environmental point of view, one of ISEW's most interesting insights is: "Efforts to control air pollution and reduce accidents have paid off by improving economic welfare during the 1970s and 1980s. . . . Improvements in both areas have had the effect of countering the generally downward trend in ISEW. They offer evidence that the choice of policies by government can indeed have a positive effect on economic welfare even if they do not increase physical output" (ibid., p. 454). However, Daly and Cobb's overall conclusion from ISEW is in stark contrast to the generally rosy GNP record: "Despite the year-to-year variations in ISEW, it indicates a long-term trend from the late 1970s to the present that is indeed bleak. Economic welfare has been deteriorating for at least a decade, largely as a result of growing income inequality, the exhaustion of resources, and the failure to invest adequately to sustain the economy in the future" (ibid., p. 455).

The diversity of Pearce et al.'s and Hueting's lists of welfare's components, the acknowledged difficulty and arbitrariness of attaching either money values or appropriate weights to them for the purpose of aggregation, and the failure of such efforts as the calculation of MEW and ISEW to generate a methodological consensus, or even a research program to generate such a consensus, raise the question as to whether in fact the derivation of a single proxy indicator for welfare is a feasible objective. Both Ruggles and Miles conclude that it is not: "No amount of imputation can convert a one-dimensional summary measure such as the GNP into an adequate or appropriate measure of social welfare" (Ruggles, 1983, pp. 41–43) and: "Despite many approaches to the modification of GNP, some of which have provided more or less useful statistical innovations, they have not resulted in, and cannot result in, any single indicator which can do for welfare or quality of life comparisons the sort of thing which GNP has achieved for economic output. . . . The search for a single indicator of progress is misguided. Social reality is too complex for it to make any sense to collapse its manifold dimensions into a one-dimensional scale" (Miles, 1992, pp. 288, 296).

This is probably a realistic conclusion. Unlike the environmental adjustments to the National Accounts which have been discussed, in order to improve their accounting of production and concerning which a considerable commonality of opinion has now been achieved, for welfare it would seem preferable to promote the adoption of a framework of indicators, of which EDP or ESNI would be one. This framework would seek to convey the dimensions of social and economic welfare more effectively than any single indicator could hope to do.

ADJUSTMENT OF THE NATIONAL ACCOUNTS

The time is ripe for the implementation of adjustments to the National Accounts for environmental factors. There is now broad agreement on the underlying principles of such adjustment and, as Lutz has said, "the pressing need is not to devise more theory or techniques, but to apply the existing methodology to concrete problems" (Lutz, 1993, p. 10).

Such adjustment could lead to an Environmentally-adjusted Domestic Product (EDP) or Environmentally Sustainable National Income

(ESNI), which, with calculations for household domestic product, could be incorporated into a broad indicator framework intended to assess socio-economic progress. Such a schema, and its possible components, are discussed in more detail in Ekins and Max-Neef (1992, Chapter 8, pp. 231-313).

Just as the methodologies and definitions underlying the present National Accounts have generated debate and dispute since their inception, so any broader indicator framework will be continually subject to criticism and possible further amendment. This is as it should be where fundamental values are concerned and where perceptions of what constitute wealth and welfare, in a broad sense, are bound to change.

As far as the value of the environment is concerned, there is, of course, much still to be learned. But much is also known and now needs to be brought formally into the accounting framework. As El Serafy (1993, p. 21) has said: "Our approach should be gradual and we should attempt to bring measurable elements into the process as our knowledge improves. But to wait until everything falls properly into place means that we shall have to wait forever."

This is an echo in an environmental context of the earlier call by Eisner: "There is more to economic activity than what is measured in conventional accounts. . . . (P)erhaps the private research has shown enough in the way of possibilities for presenting a systematic set of accounts of a greater totality. . . . It is time for the major resources of government to be put to the task. The pay-off can be great, for the economy as a whole as well as for national income accounting" (Eisner, 1988, p. 1669). Perhaps the largest payoff of all would be greater operational clarity about the achievement or lack thereof of environmentally sustainable development.

PART · THREE

*S*USTAINABLE DEVELOPMENT

CHAPTER 6

From Growth to Sustainability

On first hearing, "sustainable development" sounds difficult—and certainly achieving sustainability looks to be one of the toughest challenges our species has ever faced. But the significance of this term is simple to sum up: it indicates a switch in environmental thinking from a necessary early obsession with emerging problems to a growing interest in environmental solutions which will be economically, socially, and politically viable. Sustainable development also potentially permits a much wider range of interests (environmental, social, political, economic, commercial) to make common cause. However, there are very different views on what the end-point of this process should be, what it will involve, who is responsible for initiating and guiding it, and how fast it should go.

WHAT IS SUSTAINABLE DEVELOPMENT?

The 1992 U.N. Conference on Environment and Development, or the "Earth Summit," called for a shift from "talk" to "action". Two main categories of problem were chosen as targets. The first category is the broad range of environmental degradation which has been so thoroughly catalogued in such publications as the Worldwatch Institute's "State of the World" series. The second category embraces a range of socio-economic problems, including poverty, malnutrition, and famine. It is often argued that the first category of problems is aggravated by population and economic growth, while the second can only be alleviated by economic growth. Sustainable development aims to square the circle by proposing a new model of development which is simultaneously sustainable in environmental, economic, and social terms. But the fundamental collision between proponents of growth-based solutions and no-growth or almost-no-growth-based solutions remains at the heart of the sustainability debate.

The generally used concept of "growth", as it adheres to misleading information on economic activities which also deplete and destroy, is not suitable for any sustainable solution. "Sustainable development" is definitely meant as an alternative to growth, not as a semantic replacement.

The most commonly accepted definition of sustainable development is that first proposed in "Our Common Future:" development that meets the needs of the present without compromising the ability of future generations to meet their own needs. Fine, as far as it goes, but it can hardly be described as a blueprint. WCED Secretary-General MacNeill was a bit more specific: "An essential condition for sustainable development is that a community's and a nation's basic stock of natural capital should not decrease over time. A constant or increasing stock of natural capital is needed not only to meet the needs of present generations, but also to ensure a minimum degree of fairness and equity with future generations" (McNeill, 1990).

The concept has variously been described as a compass point, a banner, and a rallying cry. The implication is that its power to attract and direct is one of its fundamental values. "Sustainable development has become the new rallying cry of the environment/development movement," as Timberlake noted in 1988. "Like most such cries ('Liberty!', 'Power to the People!'), its meaning is not absolutely clear." By the time Pezzey (1989) settled down to review the literature in 1989, he identified around 60 separate definitions of "sustainability". In fact, there are those who see its very vagueness as part of its power to attract new recruits.

The area of thinking and discussion on sustainability is evolving rapidly. For example, in *Blueprint 3: Measuring Sustainable Development,* Pearce et al. (1994) usefully distinguish between "weak" and "strong" sustainability. The weak sustainability argument runs as follows. Any impairment of today's environment which is likely to leave tomorrow's world worse off should be compensated for. Taking a "constant capital" approach, with sustainable development defined in terms of passing on to future generations an average capital stock which is no less than that which exists now, the weak sustainability school of thought argues that we can consume or destroy environmental resources as long as we compensate for the loss by increasing the stock of manmade assets, like roads, buildings and machinery. If you adopt this perspective, the environment is seen as simply another form of capital—and, ultimately, most forms of capital are substitutable.

By contrast, the strong sustainability school introduces the concept of "critical natural capital." Examples would include the ozone layer, the carbon cycle, and biodiversity. Such resources have "primary" (i.e. ecological) values as well as "secondary" (i.e. human use-based, or market) values. Given that natural and manmade capital stocks are often not completely or directly substitutable, the strong sustainability school holds that we should at a minimum, protect critical natural capital. As "Blueprint 3" puts it: "Measuring sustainable development thus depends on the view taken about what is necessary to achieve it. On the weak sustainability rule we need to look at the overall stock of capital. On the strong sustainability rule we need to look at the overall stock of capital and pay special attention to the environment."

The intricate dynamics of sustainability have still to be adequately investigated and articulated. But already we can see enough of the outline of this new area of science, economics, and technology to understand the general direction in which society needs to travel. In this respect, the concept of sustainable development is important for three main reasons.

First, it provides a framework within which broader cultural, socio-political, economic, and technological factors can be incorporated into the environmental debate—and vice versa.

Second, the phrase is dynamic, implying a transition from unsustainable to sustainable forms of economic development and activity. It thereby potentially opens out the time-horizons within which environmental issues are discussed. The phrase introduces the notion that some problems may be more critical than others in terms of negotiating that transition. The implicit message is that we must set priorities.

Third, sustainable development is inclusive. By focusing on the need for "development," it increasingly provides a framework within which the business and development communities can feel comfortable in discussing and addressing environmental priorities. And development, clearly, is an area where business has a distinct edge on most nongovernmental organisations (NGOs)—with the result that there are growing numbers of business organizations working in this area.

THE HISTORY OF SUSTAINABLE DEVELOPMENT

When the Bretton Woods Conference was held in 1944 to establish the postwar financial and monetary system, and when the first steps to set up the SNA were made in 1945, environmental issues were not on the international political agenda. However, the last two decades, beginning with the 1972 U.N. Stockholm Conference on the Human Environment, have seen a growing number of international governments adopting environmental objectives.

In fact, probably the first use of the phrase "sustainable development" in a widely published document can be traced back to the 1980 "World Conservation Strategy," prepared by three leading international environmental agencies: IUCN (now the World Conservation Union), the United Nations Environment Programme (UNEP), and the World Wildlife Fund (WWF). This groundbreaking document led in relatively short order to the production and publication of a growing number of national sustainable development strategies or plans. It was updated in 1991 with the publication of "Caring for the Earth: A Strategy for Sustainable Living".

In addition, late in 1983, the U.N. General Assembly decided to call for the establishment of a special, independent commission—later named the World Commission on Environment and Development (WCED)—to undertake a global enquiry on the prospects of combining social and economic development with environmental protection. After three years of work, the Commission, chaired by Gro Harlem Brundtland, concluded that a transition to sustainable forms of development was possible.

The publication of the WCED report, "Our Common Future" (1989), was followed by a number of regional meetings leading up to the 1992 U.N. Conference on Environment and Development (UNCED), held in Rio de Janeiro and popularly dubbed the "Earth Summit." UNCED, which was attended by 178 national representatives, 20,000 participants and over 100 national heads of state, helped spur considerable worldwide activity. One of the main documents to emerge from UNCED, "Agenda 21," calls on the international community to promote the transition to sustainable development in developing countries.

The negotiations on the Uruguay Round of the Global Agreement on Tariffs and Trade (GATT) did not attempt to map out routes to the constraints which are linked to sustainability and trade—that task must be left to the next round. In the meantime, however, a recent "Sourcebook" prepared by the International Institute for Sustainable Development (1992) indicates the sheer range of organizations likely to be involved in such negotiations in the future.

Figure 6.1 sketches out some of the underlying trends driving the sustainable development debate. Political and regulatory pressures (trend 1) first peaked in the mid-1970s and are falling back again as the demands for deregulation and reregulation grow once again. This process is likely to be reversed, however, once it becomes clear that voluntary agreements and market mechanisms cannot deliver sustainability on their own.

Market, business-to-business and fiscal pressures (trend 2) peaked twice in the 1980s, with OPEC 1 and OPEC 2, and then shot to a third peak around 1990, with some attention focused on the potential impact of the "green consumer." In retrospect, however, 1990 marked the second highwater mark of international environmentalism—and 1994 finds us 3-4 years into the second environmental "downwave."

That said, the underlying public awareness and latent concern remains at much higher levels than hitherto. The third main trend, focused around the international drive for sustainable development, peaked first in 1987 with the WCED report, a second time shortly

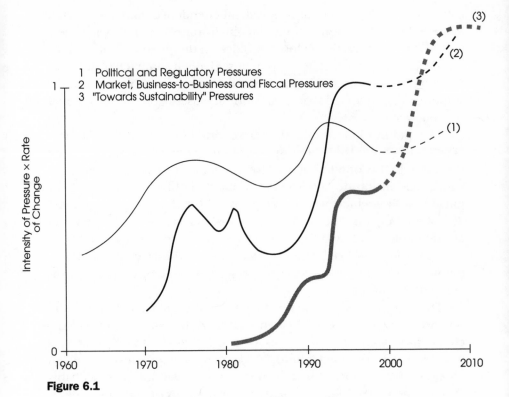

1 Political and Regulatory Pressures
2 Market, Business-to-Business and Fiscal Pressures
3 "Towards Sustainability" Pressures

Figure 6.1

before the 1992 Earth Summit, and now seems set to grow rapidly through the remaining years of the 21st century.

ACTORS

Like all revolutions, there are a considerable number of necessary conditions for the Sustainability Revolution to succeed. For example, to ensure consensus on the need for real-world progress and on specific objectives, sustainable development requires growing cooperation between different sectors of society: international government agencies, central and local government, business, environmental organisations, research and academic institutions, and individual citizens—whether acting as neighbors, voters, employees, investors, parents, or consumers. We shall briefly consider what impact the embryonic "Sustainability Revolution" has had on each of these.

Among multilateral organisations, there are five critically important institutions (Williams and Petesch, 1993):

THE U.N. COMMISSION ON SUSTAINABLE DEVELOPMENT (CSD)

As the main institutional product of UNCED, the CSD's mission is to raise the resources needed to translate Agenda 21 into effective international and domestic policies and initiatives. Guided by an annual thematic agenda, the CSD is seen as the international policy arena for debate on the world's emerging priorities in this area—and as coordinator of the programmes of the various international development institutions. The CSD has set itself an ambitious program, in the spirit of Agenda 21, but it remains to be seen whether it can deliver in any sensible time-scale.

THE U.N. DEVELOPMENT PROGRAMME (UNDP)

UNDP has developed its own program for achieving the technical and technology transfers needed to promote sustainable development in the developing world. "Capacity 21", which originated from UNCED, is designed to focus UNDP aid in such a way as to help interested developing countries to produce and adopt their own environmentally-sound development (or, to use the UNDP term, sustainable human development) strategies and action plans. Increased public participation is a central objective. In addition, UNDP has launched a Sustainable Development Network (SDN) to facilitate access by developing countries to information on key issues associated with sustainable development.

UNDP is well positioned to make a major contribution to putting sustainable development into practice, but faces financial constraints which raise real questions about its own long-term capacities.

THE U.N. ENVIRONMENT PROGRAM (UNEP)

UNEP's main achievements to date have been in terms of environmental monitoring and, as in the case of the Montreal Protocol, of extending the framework of international law to cover emerging environmental problems. The agency needs considerable strengthening to ensure that this mainstream activity is carried out effectively. Potentially, it could also play a central role in organizing periodic reviews of sustainable development progress and needs in different world regions, for the CSD and the U.N. regional commissions for economic and social development. UNEP's Paris-based Industry and Environment Program Activity Center (IE/PAC) is involved in highly relevant work in such areas as cleaner technologies, life-cycle assessment and corporate environmental reporting.

THE WORLD BANK

Long the ogre of the international environmental movement, the World Bank has been trying increasingly hard to "green" its thinking and operations. In 1987, the Bank adopted environmental policies and guidelines as part of a new strategy. This strategy is based on the assumption that economic growth and environmental protection should somehow proceed together and that the Bank needs to focus on institutional capacity building quite as much as on mainstream investment lending (World Bank, 1993).

Inevitably, the Bank has been somewhat schizophrenic in relation to the sustainable development agenda, but an important signal of its willingness to change was its establishment of an internal Vice Presidency for Environmentally Sustainable Development (ESD). A central challenge for the Bank will be the area of "green conditionality," which would ensure that sustainable development conditions and projects are accompanied by packages of related training, to ensure that recipients have a real chance of achieving sustainable development. (For an elaboration of the World Bank view on sustainable development, see Chapter 7 by Serageldin, Daly, and Goodland.)

THE GLOBAL ENVIRONMENTAL FACILITY (GEF)

Launched in 1991 by the World Bank, UNDP, and UNEP, the GEF aims to supply concessionary funding for projects with significant global environmental benefits. Still largely at the pilot phase, the GEF will need strong international backing if it is to achieve its potential—particularly now its remit has been expanded under the terms of the U.N. Conventions on Climate Change and Biodiversity.

OTHER INTERNATIONAL ORGANIZATIONS

A considerable number of other international public sector organizations are involved in sustainable development programs. They include the Organization for Economic Cooperation and Development (OECD), financial institutions (e.g. the European Bank for Reconstruction and Development (EBRD) and the Asian Development Bank (ADB, 1993), and regional government agencies such as the European Commission, which published its Fifth Environmental Action Programme, entitled "Towards Sustainability," in 1992.

Overall, however, the international government and financial systems have scarcely even begun to play with the concept of weak sustainability. Implementing even this partial agenda will be a massive challenge for today's institutions. If circumstances dictate that the world must travel the strong sustainability route, there is no question but that such institutions will need a radical overhaul—and totally new institutions will be required to manage the global and regional "commons."

The 1992 Green World Survey (Elkington and Dimmock, 1992) identified national government sustainable development programs in many of the countries covered and new initiatives have continued to be announced. In Canada, for example, a series of federal and regional roundtables have brought together different sectors of society to develop a consensus on priorities and necessary actions. In the United States, there is now a President's Council on Sustainable Development. And Japan has produced a 100-year plan, "New Earth 21" (Fornander, 1991), covering, in broad outline, some of the key tasks that need to be addressed in moving towards sustainability. It lays out some key steps for Japan, Inc. in pursuing sustainability. For the first five decades, the steps are as follows: 1990–2000, "The New Earth Energy Conservation Program"; 2001–2010, "The New Earth Clean Energy Program"; 2011–2020, "The New Earth Environment-Friendly Technology Program"; 2021–2030, "The New Earth Carbon Dioxide Sinks Program"; and 2031–2040, "The New Earth Carbon Dioxide Sinks Program".

Individual EU member states have been preparing their own strategies and programs, although their approaches and levels of commitment vary widely. So, for example: Denmark published its governmental Action Plan in December 1988 (State Information Service, 1988); The Netherlands published its National Environmental Policy Plan (NMP) in 1989, NMP-Plus in 1991 and NMP-2 in 1994; and the

United Kingdom "This Common Inheritance" in 1990, the first year report in 1991, the second in 1992, and the "UK Strategy for Sustainable Development" in 1994. A number of the potential new members of the European Union have also been developing their strategies in this field: the Swedish Environmental Protection Agency (SNV), for example, presented its new action plan, "Strategy for Sustainable Development," in 1993.

Of these European initiatives, the Dutch is by far the most far-reaching—and has significantly influenced the thinking of the European Commission. In support of its NMP-style strategic work, The Netherlands has also been pioneering other routes to reducing the environmental burden caused by the country's industrial and other activities. A key feature of the process has been the introduction of so-called "environmental covenants," establishing agreed-upon targets for improvements in the environmental performance of key industrial sectors.

Despite these anecdotal examples of progress, the sustainable development process has only just begun; and, in all probability, will evolve over decades rather than years. Hopefully, however, the concept of sustainability is also now beginning to reach parts of the global village which still have an opportunity to build a sustainable economy from the outset. The Seychelles, for example, are trying to strike a new balance with tourism with a tough environmental plan—"Environmental Management Plan for the Seychelles 1990-2000" (Hanneberg, 1991)—which adopts sustainable development as a guiding principle.

But we are still a long way from the day when all of the world's 180-odd nation states have even a first generation sustainability development plan in place. Given the pace of environmental degradation, this can only be described as a case of "too little, too late." The role of local, national, and international NGOs is therefore vitally important in keeping both the debate on the boil and the sustainability locomotive steaming down the track.

NGOs have played—and will continue to play—a central role in the process of defining, committing to and implementing sustainable development. Indeed, the sustainable development movement was catalysed by two key NGOs, IUCN (now the World Conservation Union) and WWF (then the World Wildlife Fund; now, in some countries, the World Wide Fund For Nature). The number of NGOs interested, or active, in this field has exploded. The Center for Our Common Future, set up in 1987 to coordinate NGO inputs to the implementation of, and followup on, Our Common Future, currently lists 60,000 organisations on its database, a high proportion of which are NGOs. India alone is said to have over 9,000 NGOs. Such organizations are growing rapidly in newly-democratized parts of the world such as Central and Eastern Europe and the former Soviet Union. NGOs are now mapping out the transition to a more sustainable future: the World Resources Institute (WRI, 1993), to take just one example, is working on "The 2050 Project" alongside the Santa Fe Institute and The Brookings Institution. The 4-year-project is exploring the approaches and policies needed to achieve a sustainable global society by the middle of the next century. It is focusing on the environmental, economic, and social dimensions of sustainability and on the transitions needed to achieve them.

The sustainable development credo is not one that appeals to all— or perhaps even most—NGOs. One of the key architects of the Brundtland Commission formulation of sustainability, Secretary General Jim MacNeill (1990), has said that: "The maxim for sustainable development is not limits to growth; it is the growth of limits." This definition, based on stretching environmental limits to accommodate expanding human needs, will certainly not appeal to those who argue for a "steady-state economy," let alone to supporters of the "deep ecology" worldview.[1] But the mainstream sustainable development community recognizes that we are faced with some fairly inexorable population trends. Faced with this challenge, the argument runs, the imaginative use of emerging technologies can help relax "natural" limits by boosting our resource efficiency, that is, helping us do more with less. In the end, however, limits will be definite, and the steady state unavoidable.

Although the liable and legalistic part of the international business community is embarking on the Sustainability trend, at the very same time, critics of industry can point to exactly opposite movements. Critics of the North American Free Trade Agreement (NAFTA) note that hundreds of US companies have been joining the rush to capitalise on Mexico's cheap labour and lax environmental laws by locating along the 1,250-mile border between Mexico and the United States. More than 2,000 such companies, or "maquilas," now represent Mexico's second largest source of foreign exchange after oil and are help-

---

[1] See "Deep Ecology", pp. 242–247 in Dobson (1991).

ing to create environmental legacies which will haunt Mexico for decades to come.

But the concept of sustainable development is rarely now seen as evidence of an imperialist or neo-colonialist conspiracy. Growing numbers of newly industrialized and developing countries recognize that this has to be the way forward—the real issue is how fast to move and who pays. It will be well worth watching countries like India over the next decade: the Confederation of Indian Industry (CII, 1992) is working hard to disseminate to Indian Companies the thinking of organizations like the Business Council for Sustainable Development.

Business has potentially a great deal to bring to the process of sustainable development, not least in terms of prioritization processes, the management of resources, finance, and technology. In addition, some pioneering companies, as part of their corporate reporting (SustainAbility, 1993a; and Elkington and Robins, 1994) and new product development or eco-efficiency activities, are developing innovative approaches to life-cycle assessment (SustainAbility, 1993b), lifecycle design, and full cost accounting.

The banking and investment communities still have a long way to go in terms of incorporating environmental accounting principles and sustainability indicators into their financial equations. The fact that some 20 percent of the recent catastrophic losses incurred by Lloyds insurance syndicates were related to environmental risks (e.g. asbestos and the disposal of toxic and radioactive wastes) may help to wake up the financial world to the threats currently building, and, interestingly, Greenpeace has targeted the insurance community with its work on the risks associated with climate change. But this is an area where the process of change is still embryonic at best.

More fundamental still, the sustainability implication which few, if any, companies have yet considered relates to the issue of "need." Markets grow by inventing new needs, new wants, new desires. At some point, the sustainable development vector will intersect with this aspect of the market economy, with the result that business will increasingly find that it is no longer enough simply to consider the environmental performance of a product from cradle to grave, but, even more fundamentally, we will have to ask whether the likely impacts are balanced by the meeting of real needs.

From analyzing the effects of toxic chemicals on human and animal immune systems through work on ozone depletion and global warming to research on biodiversity, the global science community has been a key actor in the environmental revolution. The need for objective, yet farsighted scientific inputs has never been greater, as society increasingly has to prioritize risks before allocating scarce financial and other resources. At the same time, vast amounts of research and development effort are needed to begin the transition to more sustainable forms of natural resource utilization (Rosenberg et al., 1993), industrial production, consumption, urban development, and transport. Alongside this scientific work, the "dismal science" of economics needs to be re-invented. The work of groups like the New Economics Foundation (NEF) is already mapping ways forward.

Ultimately, sustainable development has to be sold to the citizens of every country in the world, whether they are acting as industry's neighbors, as employees, voters, parents, or consumers. Despite public opinion polls showing surprisingly high levels of environmental awareness, concern, and action in a wide range of countries (Dunlap, Gallup, and Gallup, 1992), the public's support cannot be taken for granted. These are the people who will ultimately have to pay and to sacrifice other possible rewards to ensure that future generations inherit a world worth living in. We need to see more advertising and media campaigns which carry aspects of the sustainability message through to the general public in their everyday lives—standing in bus shelters or waiting for trains.

The critical role of ordinary citizens is recognised in the European Commission's Fifth Environmental Action Programme—and the European Union may well be worth watching as it struggles with the task of communicating sustainability to an increasingly diverse range of nationalities. This, in microcosm, will represent a useful model of the challenge facing the world as a whole.

REAL-WORLD SOLUTIONS

Sustainable development is a process which necessarily involves all sectors of society, including those not yet born. The early phase—when key actors in this process felt they faced "no win" (where no one could win) or "zero sum" (where only one side could win) outcomes—has been replaced by a phase in which key actors, including NGOs, have been looking for "win-win" outcomes, where both sides can benefit. The central challenge for the future is to work out how to achieve "win-win-win" outcomes, where both sides *and* the environment benefit. This by no means implies that conflict should be avoided, on the contrary: conflict is a function of the great choices ahead.

Sustainable development principles are beginning to be adopted, or at least experimented with, by some international government agencies, by some national governments and by some corporations and business organisations. Unfortunately, however, the early involvement of the business community (and particularly of the marketing profession) has spurred the process whereby the words "sustainable" and "sustainability"—like "green" or "environment-friendly" before them—are strapped onto other words. So one sees phrases like "sustainable use" (of natural resources), "sustainable society" (or economy), "design for sustainability," "sustainable manufacturing," "sustainable mobility," "sustainable tourism," "sustainable futures," and even "sustainable growth." This process seems to simply give a green light for further consumption, further production, and further growth, which is a falsification. But, from another perspective, this can be seen as an important first stage in the selling of the idea of sustainability to the citizens—and to voters, who will need to understand why real sacrifices may be needed to ensure a genuinely sustainable future.

If there is one thing most long-standing proponents of sustainable development would agree upon, it is this: to arrive at a sustainable state, our industrialized societies will have to re-invent themselves, just as agricultural societies re-invented themselves during the Industrial Revolution. As a result, Meadows et al. (1992) have spoken in terms of the "Sustainability Revolution." But sustainable development will only move from the realm of talk to the world of action if it is seen to offer real-world solutions to recognized problems.

Anyone searching for a visionary account of what all of this could mean in terms of everyday realities for governments, corporations and citizens would be well advised to read Hawken's excellent book, *The Ecology of Commerce* (1993). In talking of the need to make the transition to a "restorative economy," Hawken sketches the extraordinary scale of the challenge now facing us. Reassuringly, he also shows that the transition is still likely to be manageable and could well turn out to be one of the most exciting historical stages which our species has yet gone through. At long last, we seem to be approaching the point where the list of possible solutions is beginning to catch up with the still growing list of problems.

CHAPTER 7

The Concept of Sustainability

The paramount importance of sustainability arose partly because the world is recognizing that current patterns of economic development are not generalizable. Present patterns of OECD per capita resource consumption and pollution cannot possibly be generalized to all currently living people, much less to future generations, without liquidating the natural capital on which future economic activity depends. Sustainability thus arose from the recognition that the profligate and inequitable nature of current patterns of development, when projected into the not too distant future, lead to biophysical impossibilities. The transition to sustainability is urgent because the deterioration of global life-support systems that is, the environment, imposes a time limit. We do not have time to dream of creating more living space or more environment, such as colonizing the moon or building cities beneath the sea; we must save the remnants of the only environment we have, and allow time for, and invest in, the regeneration of what we have already damaged.

NATURAL CAPITAL AND SUSTAINABILITY

Sustainability in economic terms can be described as the "maintenance of capital", sometimes phrased as "non-declining capital". Historically, at least as early as the Middle Ages, the merchant traders used the word "capital" to refer to human-made capital. The merchants wanted to know how much of their trading ships cargo sales receipts could be consumed by their families without depleting their capital. Of the forms of capital (discussed in Figure 7.1) environmental sustainability refers to natural capital. So the definition of environmental sustainability includes at least two further terms, namely "natural capital," and "maintenance" (or at least "non-declining").

Natural capital is basically our natural environment, and is defined as the stock of environmentally provided assets (such as soil, atmosphere, forests, water, wetlands) which provide a flow of useful goods or services. The flow of useful goods and services from natural capital can be renewable or nonrenewable, and marketed or nonmarketed. Sustainability means maintaining environmental assets, or at least not depleting them. "Income" is sustainable by the generally accepted Hicksian definition. Any consumption that is based on the depletion of natural capital should not be counted as income. Prevailing models of economic analysis tend to treat consumption of natural capital as income, and therefore tend to promote patterns of economic activity that are unsustainable. Consumption of natural capital is liquidation, the opposite of capital accumulation.

Natural capital is distinguished from other forms of capital, namely human capital or social capital (people, their capacity levels, institutions, cultural cohesion, education, information, knowledge),[1] and hu-

---

[1] Human capital formation, by convention is left out of the national accounts for various reasons, one of which is that, if it is truly productive, it will eventually be reflected, through enhanced productivity, in a higher GDP. Realization of the values of education and administration, for example, are lagged, and are convention-

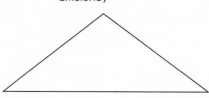

Economic objectives:

- Growth
- Equity
- Efficiency

Social objectives:

- Empowerment
- Participation
- Social mobility
- Social cohesion
- Cultural identity
- Institutional
  development

Ecological objectives:

- Ecosystem integrity
- Carrying capacity
- Biodiversity
- Global issues

Social Sustainability ("SS")

ES needs SS—the social scaffolding of organizations that empower self-control and self-policing in peoples' management of natural resources (see Cernea, 1993). Resources should be used in ways which increase equity and social justice, while reducing social disruptions. SS will emphasize qualitative improvement over quantitative growth; and cradle-to-grave pricing to cover full costs, especially social. SS will be achieved only by strong and systematic community participation or civil society (Putnam, 1993 a, b). Social cohesion, cultural identity, institutions, love, commonly accepted standards of honesty, laws, discipline, etc., constitute the part of social capital that is least subject to measurement, but probably most important for SS. This "moral capital," as some have called it, requires maintenance and replenishment by the religious and cultural life of the community. Without this care it will depreciate as surely as will physical capital.

Economic Sustainability ("EcS")

The widely accepted definition of economic sustainability is "maintenance of capital," or keeping capital intact, and has been used by accountants since the Middle Ages to enable merchant traders to know how much of their sales receipts they and their families could consume. Thus the modern definition of income (Hicks, 1946) is already sustainable. But of the four forms of capital (human-made, natural, social, and human) economists have scarcely at all been concerned with natural capital (e.g., intact forests, healthy air) because until relatively recently it had not been scarce. Also, economists prefer to value things in money terms, so we are having major problems valuing natural capital, intangible, intergenerational, and especially common access resources, such as air etc. In addition, environmental costs used to be "externalized," but are now starting to be internalized through sound environmental policies and valuation techniques. Because people and irreversibles are at stake, economics has to use anticipation and the precautionary principle routinely, and should err on the side of caution in the face of uncertainty and risk. Human capital (investments in education, health, and nutrition of individuals) is now accepted in the economic lifestyle (WDR, 1990, 1991, 1995), but social capital, as used in SS, is not adequately addressed.

Environmental Sustainability ("ES")

Although environmental sustainability is needed by humans and originated because of social concerns, ES itself seeks to improve human welfare by protecting the sources of raw materials used for human needs and ensuring that the sinks for human wastes are not exceeded, in order to prevent harm to humans. Humanity must learn to live within the limitations of the physical environment, both as a provider of inputs ("sources") and as a "sink" for wastes (Serageldin, 1993a). This translates into holding waste emissions within the assimilative capacity of the environment without impairing it and by keeping harvest rates of renewables within regeneration rates. Quasi-ES can be approached for non-renewables by holding depletion rates equal to the rate at which renewable substitutes can be created (El Serafy, 1991).

Figure 7.1   Comparison of social, economic, and environmental sustainability (from: Serageldin 1993a, b)

NOTE: Clearly these three concepts are related closely in parts. However, this chapter focuses on environmental sustainability, rather than on sustainable development. The moment "development" is introduced, the discussion becomes quite different, and thus is left for another occasion.

man-made capital (houses, roads, factories, ships). For the mercantilists, until very recently capital referred to the form of capital in the shortest supply, namely human-made capital. Investments were made in the limiting factor, such as sawmills and fishing boats, because their natural capital complements—forests and fish—were abundant. That idyllic era has ended.

Now that the environment is so heavily used, natural capital has become the limiting factor for much economic development, as much as human-made capital. In some cases, like marine fishing, it has become *the* limiting factor. Fish have become limiting, rather than fishing boats. Timber is limited by remaining forests, not by saw mills; petroleum is limited by geological deposits and atmospheric capacity to absorb CO_2, not by refining capacity. As natural forests and fish populations become limiting we begin to invest in plantation forests and fish ponds. This introduces a hybrid category that combines natural and human-made capital—a category we may call "cultivated natural capital".[2] This category is vital to human well-being, accounting for most of the food we eat, and a good deal of the wood and fibers we use. The fact that humanity has the capacity to "cultivate" natural capital dramatically expands the capacity of natural capital to deliver services, but does not avoid entirely the limiting role of such capital.

NATURAL CAPITAL IS NOW SCARCE

In an era in which natural capital was considered infinite relative to the scale of human use, it was reasonable not to deduct natural capital consumption from gross receipts in calculating income. That era is now past. The goal of environmental sustainability is thus the conservative effort to maintain the traditional meaning and measure of income in an era in which natural capital is no longer a free good, but is more and more the limiting factor in development. The difficulties

---

ally assumed to be equal to their costs. The loss of natural capital, if not recorded, as is largely the case today, may take some time before it will be reflected in income and productivity measurements.

[2] The subcategory of marketed natural capital, intermediate between human capital and natural capital, is "cultivated natural capital," such as agriculture products, pond-bred fish, cattle herds, and plantation forests.

in applying the concept arise mainly from operational problems of measurement and valuation of natural capital, as emphasized in other chapters of this report, and by Ahmad et al. (1989), Lutz (1993), and El Serafy (1991, 1993).

DEFINITIONS OF ENVIRONMENTAL SUSTAINABILITY

Sustainability has several levels—weak, sensible, strong, and absurdly strong—depending on how strictly one elects to hew to the concept of maintenance (or non-declining) capital (Daly and Cobb, 1989). We recognize that there are at least four kinds of capital: Human-made (the one usually considered in financial and economic accounts), natural capital (as defined previously, and leaving for the moment the case of cultivated natural capital), human (investments in education, health and nutrition of individuals), and social (the institutional and cultural basis for a society to function).

Weak sustainability is maintaining total capital intact without regard to the composition of that capital among the different kinds of capital (natural, man-made, social, or human). This implies that the different kinds of capital are perfect substitutes, at least within the boundaries of current levels of economic activity and resource endowment.

Sensible sustainability would require that, in addition to maintaining the total level of capital intact, some concern should be given to the composition of that capital between natural, human-made, human, and social. Thus, oil may be depleted as long as the receipts are invested in other capital (e.g., human capital development) elsewhere; but, in addition, efforts should be made to define critical levels of each type of capital, beyond which concerns about substitutability could arise and these should be monitored to ensure that the patterns of development do not promote a total decimation of one kind of capital no matter what is being accumulated in the other forms of capital. This assumes that while human-made and natural capital are substitutable over a sometimes significant but limited margin, they are complementary beyond that limited margin. The full functioning of the system requires at least a mix of the different kinds of capital. Since we do not know exactly where the boundaries of these critical limits for each type of capital lie, it would behoove the sensible person to err on the side of caution in depleting resources (especially natural capital) at too fast a rate.

Strong sustainability requires maintaining different kinds of capital intact separately. Thus, for natural capital, receipts from depleting oil should be invested in ensuring that energy will be available to future generations at least as plentifully as enjoyed by the beneficiaries of today's oil consumption. This assumes that natural and human-made capital are not really substitutes but complements in most production functions. A saw-mill (human-made capital) is worthless without the complementary natural capital of a forest. The same logic would argue that if there are to be reductions in one kind of educational investments they should be offset by other kinds of education, not by investments in roads.

Absurdly strong sustainability would never deplete anything. Nonrenewable resources—absurdly—could not be used at all; for renewables, only net annual growth rates could be harvested, in the form of the overmature portion of the stock.

The schematic presentation above, taken from Serageldin and Steer (1994a) highlights the general direction to be taken for this immediate discussion, which can usefully be limited for the time being to the perceived tradeoffs between human-made capital and natural capital, fully recognizing that the issues are more complex than can be fully explored in the following paragraphs.

Economic logic requires us to invest in the limiting factor which now includes natural capital as much as human-made capital, which was limiting in the past. Investing in natural capital (nonmarketed) is essentially an infrastructure investment on a grand scale, that is the biophysical infrastructure of the entire human niche. Investment in such "infra-infrastructure" maintains the productivity of all previous economic investments in human-made capital (public or private) by rebuilding the natural capital stocks that have come to be limitative. Operationally, this translates into the following:

- encouraging the growth of natural capital by reducing our level of current exploitation;
- investing in projects to relieve pressure on natural capital stocks by expanding cultivated natural capital, such as tree plantations to relieve pressure on natural forests;
- and increasing the end-use efficiency of products (improved cookstoves, solar cookers, hay-box cookers, wind pumps, solar pumps, manure rather than chemical fertilizer, etc.).

CRITERIA FOR ENVIRONMENTAL SUSTAINABILITY

From the above "maintenance of natural capital" approach to environmental sustainability (ES), we can draw practical rules-of-thumb (Figure 7.2) to guide the design of economic development. As a first approximation, the design of investment strategies should be compared with the input/output rules as in Figure 7.2, in order to assess the extent to which a project is sustainable. At the next level of detail, specific indicators of environmental sustainability (such as those the World Bank and others are preparing) can be used.

The implications of implementing environmental sustainability are immense. We must learn how to manage the renewable resources for the long term; we have to reduce waste and pollution; we must learn how to use energy and materials with scrupulous efficiency; we must learn how to use solar energy in all its forms; and we must invest in repairing, as much as possible, the damage done to the earth in the past few decades by unthinking industrialization in many parts of the globe. Environmental sustainability needs enabling conditions which are not integral parts of environmental sustainability: not only economic and social sustainability (Figure 7.1) but democracy, human resource development, empowerment of women, and much more investment in human capital than is common today (i.e., increased literacy, especially ecoliteracy) (Orr, 1992).

The sooner we start to approach environmental sustainability the easier it will become. For example, the demographic transition took

---

1. Output Rule:
   Waste emissions from a project should be within the assimilative capacity of the local environment to absorb without unacceptable degradation of its future waste absorptive capacity or other important services.

2. Input Rule:
   (a) Renewables: harvest rates of renewable resource inputs would be within regenerative capacity of the natural system that generates them.
   (b) Non-renewables: depletion rates of non-renewable resource inputs should be equal to the rate at which renewable substitutes are developed by human invention and investment. Part of the proceeds from liquidating non-renewables should be allocated to research in pursuit of sustainable substitutes. *

Figure 7.2 Rules-of-thumb for environmental sustainability (from Serageldin, 1993a, b)

* For a theoretical development of this idea, see El Serafy (1991, 1993) and Dasgupta and Heal (1979).

a century in Europe, but only a decade in Taiwan: technology and education make big differences. But the longer we delay, the worse the eventual quality of life (e.g., fewer choices, fewer species, more risk), especially for the poor who do not have the means to insulate themselves from the negative effects of environmental degradation.

Many writers have expressed concern that at present the world is hurtling away from environmental sustainability (Simonis, 1990; Meadows et al., 1992, Brown et al., 1994, Hardin, 1993), although consensus has not yet been reached. But what is not contestable is that the modes of production prevailing in most parts of the global economy are causing the exhaustion and dispersion of a one-time inheritance of natural capital, such as topsoil, groundwater, tropical forests, fisheries, and biodiversity (see Chapter 4).

Yet what is galling is that in spite of spending capital inheritance rather than just income, most of the world consumes at barely subsistence levels. Can humanity attain a more equitable standard of living which does not exceed the carrying capacity of the planet? The transition to environmental sustainability will inevitably occur. However, it remains in doubt as to whether nations will have the wisdom and foresight to plan for an orderly and equitable transition to environmental sustainability, rather than allowing biophysical limits to dictate the timing and course of this transition.

It is obvious that if pollution and environmental degradation were to grow at the same rate as economic activity, or even population growth, the damage to ecological and human health would be appalling, and the growth itself would be undermined and even self-defeating. However, a transition to sustainability is possible, although it will require changes in policies and the way humans value things. The key to the improvement of the well-being of millions of people lies in the increase of the added value of output after properly netting out all the environmental costs and benefits and after differentiating between the stock-and-flow aspects of the use of natural resources. Without this necessary adjustment in thinking and measurement, pursuit of economic growth that does not account for natural capital, and which counts depletion of natural capital as an income stream will *not* lead to a sustainable development path. The global ecosystem, which is the source of all the resources needed for the economic subsystem, is finite and has now reached a stage where its regenerative and assimilative capacities have become very strained. It looks inevitable that the next century will witness a doubling of the number of people in the human economy, depleting sources and filling sinks with their increasing wastes. If we emphasize the latter, it is because human experience seems to indicate that we have tended to overestimate the environment's capacity to cope with our wastes, even more than we overestimated the "limitless" bounty of such resources as the fish in the sea.

A single measure—population times per capita consumption of natural capital—encapsulates an essential dimension of the relationship between economic activity and environmental sustainability. This scale of the growing human economic subsystem is judged, whether large or small, relative to the finite global ecosystem on which it so totally depends, and of which it is a part. The global ecosystem is the source of all material inputs feeding the economic subsystem, and is the sink for all its wastes. Population times per capita consumption of natural capital is the total flow (throughput) of resources from the global ecosystem to the economic subsystem, then back to the global ecosystem as waste (as dramatized in Figure 7.3). In the long gone "empty world" case, the scale of the human economic subsystem is small, relative to the large, but non-growing global ecosystem. In the lower diagram, the "full world" case in 7.3C, the scale of the human economic subsystem is large and still growing, relative to the finite global ecosystem. In the full world case, the economic subsystem has already started to interfere with global ecosystemic processes, for instance, altering the composition of the atmosphere, or the now nearly global damage to the ozone shield.

As we said in Chapter 4, carrying capacity is a measure of the amount of renewable resources in the environment in units of the number of organisms these resources can support. It is thus a function of the area and the organism: a given area could support more lizards than birds with the same body mass. Carrying capacity is difficult to estimate for humans because of major differences in affluence, behavior, and technology. An undesirable "factory-farm" approach could support a large human population at the lowest standards of living: certainly the maximum number of people is not the optimum. The higher the throughput of matter and energy, or the higher the consumption of environmental sources and sinks, the smaller the number of people that can enjoy it.

Ehrlich and Holdren (1974) encapsulate the basic elements of this concept in a simple and forceful, though static, presentation: The impact (I) of any population or nation upon environmental sources and sinks is a product of its population (P), its level of affluence (A), and

a) The economy as an isolated system

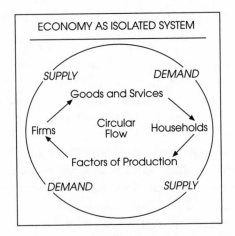

b) Linking the economic and environmental systems

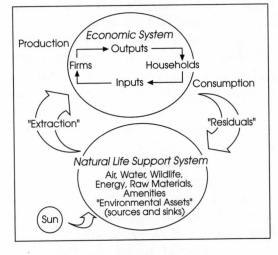

c) The economy dependent on the environment

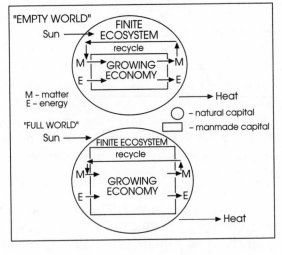

Figure 7.3 Relationship between the economic and the environmental systems: three views

the damage done by the particular technologies (T) that support that affluence.[3]

There are a number of ways of reducing the impacts of human activities upon the environment. These include changing the structure of production and demand (i.e. more high-value, low-throughput production and service industries) and investing in environmental protection (e.g. for the amenity value, if for nothing else). But for the sake of simplicity, we will first proceed with the static analysis of the Ehrlich and Holdren formulation before we move on to add a more dynamic understanding of sustainability issues. As a start, we will look at reducing environmental impacts of human activities upon the environment through the three variables in the equation, namely: limiting population growth; limiting affluence; and improving technology, thereby reducing throughput intensity of production. There is much to be done to limit the impact of human activities upon the environment, although so far, many of the measures have proven politically unpopular and difficult to achieve. The changes in variables—population, affluence, technology—through which the impact can be limited are each examined in more detail below.

POPULATION

Population stability is fundamental to environmental sustainability. Today's population of 5.5 billion people is increasing by nearly 100 million a year. Just the basic maintenance of 100 million extra people per year demands an irreducible minimum of throughput in the form of clothing, housing, food, and fuel. There is so much momentum in population growth that even under the United Nations most optimistic scenario, the world's population may finally level off at 11.6 billion in 2150! Since under current inequitable patterns of production, consumption, and distribution, we have not provided adequately for one-fifth of humanity at today's relatively low population, the prospects

---

[3] $I = P \times A \times T$
$[I = P \times Y/P \times I/Y]$

Population (P) refers to human numbers.
Affluence (Y/P) is output (Y) per capita.
Technology (I/Y) refers to environmental impact per unit of output, i.e., a dollar's worth of solar heating stresses the environment less than a dollar's worth of heat from a lignite-fired thermal power plant.

for being either able or willing to provide better for double that number of people look grim indeed, unless major changes in attitudes and practices were to happen. We do not want to cast a political problem (willingness to share) as a biophysical problem (encountering limits to total product). We urge much greater sharing. However, we do not want to make the opposite error of suggesting that more equitable sharing will permit us to avoid completely the issue of biophysical limits to total production in the face of mounting population pressure. Responsible stewardship of the earth requires that we redouble efforts to slow down population growth, especially in the poorest and most vulnerable countries, where population is currently growing fastest and people are suffering most.

The human population is totally dependent on energy from the sun, fixed by green plants, for all food, practically all fiber (cotton, wool, paper), most building materials (wood), and most of the cooking and heating fuels in many developing countries (fuelwood). The human economic subsystem now appropriates 40 percent of net terrestrial photosynthesis, according to Vitousek et al. (1986). Yet the sun provides enough energy to cover 6,500 times the total commercial energy consumption of the world. Instead of harnessing this massive source of clean and renewable energy, the bulk of energy research funds are still going to nuclear energy. This speaks poorly of the priorities set for energy research worldwide and is a measure of how far we still have to go to get the concept of sustainability thoroughly incorporated into the priorities of those policymakers allocating the energy research dollars.

Whether the issue will be joined over the energy fixed by photosynthesis or not, there are reasons to be concerned. Several factors are all working in the same direction to reduce irreversibly the energy available globally through plants. Greenhouse warming, damage to the ozone shield, and unstable climates all seem inescapable and may have started. Depending on the models used, these will reduce agricultural, forest, fisheries, rangeland, and other yields. The increases in ultraviolet-b light reaching the earth through the damaged ozone shield may decrease the carbon-fixing rates of marine plankton, one of the biggest current carbon sinks. In addition, ultraviolet-b light may damage young or germinating crops. According to some reports, tiny temperature elevations have already begun to increase the decomposition rates of the vast global deposits of peats, soil organic matter, and muskeg, thus releasing stored carbon. Only in mid-1992 did the circumboreal muskeg and tundra become net global carbon sources (instead of being net C-sinks). Some claim that at least an immediate 50 percent reduction in global fossil fuel use is necessary to stabilize atmospheric composition. Whether one accepts this estimate or not, it dramatizes the gravity of the situation. There really is no ground for complacency.

AFFLUENCE

Overconsumption by the OECD countries contributes more to some forms of global unsustainability than does population growth in low income countries (Mies, 1991; Parikh and Parikh, 1991). If energy consumption is used as a crude surrogate for environmental impact on the earth's life support systems, (crude since the type of energy used is not taken into account), then "A baby born in the United States represents twice the impact on the Earth as one born in Sweden, three times one born in Italy, 13 times one born in Brazil, 35 times one in India, 140 times one born in Bangladesh or Kenya, and 280 times one born in Chad, Rwanda, Haiti, or Nepal" (Ehrlich and Ehrlich, 1989a, b). Although Switzerland, Japan, and Scandinavia, for example, have recently made great progress in reducing the energy intensity of production, the key question is: can humans lower their per-capita impact (mainly in OECD countries) at a rate sufficiently high to counterbalance their explosive increases in population (mainly in low-income countries)? The affluent are reluctant to acknowledge the concept of sufficiency, and thus, to begin emphasizing quality and non-material satisfactions. Redistribution from rich to poor on any significant scale is, at present, felt to be politically impossible. But the questions of increasing equity in sharing the earth's resources and its bounty must be forcefully put on the table.

Increased affluence, especially of the poor, thus need not inevitably hurt the environment. Indeed, used wisely, economic growth can provide the resources needed to protect and enhance the environment in the poorest developing countries where environmental damage is caused as much by the lack of resources as it is by rapid industrialization. Indeed if Africa is to have any hope of protecting its forests and soils, accelerated economic and human development is imperative.

Thus, there is a nexus of problems linking poverty, environmental degradation, and rapid population growth. Breaking this nexus of problems is essential if the poor are not to continue to be the victims, as well as the unwitting cause, of environmental degradation.

Accelerate the two crucial transitions: to population stability, and to renewable energy. The main reasons are:

1) Human capital formation: education and training, employment creation, particularly employment for girls equivalent to that for boys; meeting unmet family planning demand.

2) Technological transfer: for the South and East to leapfrog the North's environmentally damaging stage of economic evolution. For developing countries, this requires creating an incentive framework conducive to efficient investment. For industrial countries, this requires an open trade regime and adequate investment in new, cleaner technologies.

3) Direct poverty alleviation: including social safety nets, and targeted aid (inter alia see World Development Report 1990, Goodland and Daly, 1993a, b).

Figure 7.4   Priorities to approach environmental sustainability

TECHNOLOGY

Technology continues to play a vital role in driving a wedge between economic activity and environmental damage. Illustrations of this occur in virtually every field of human activity.

In energy, for example, the introduction of mechanical and electrical devices in power generation over the past four decades has reduced particulate emissions per unit of energy generated by up to 99 percent; and newer technologies, such as flue gas desulfurization and fluidized bed combustion are dramatically reducing emissions of sulfur and nitrogen oxides. But it will be the transition to nonfossil-based sources that will make the permanent difference. Here, technological progress has been remarkable—with costs of solar generation of electricity falling by 95 percent in the past two decades—but it is not yet enough. Renewable energy continues to receive much too small a share of public research funds.

Technological innovation and application has also done much to make agriculture more sustainable. New technologies have enabled a doubling of food production in the world in just 25 years, with more than 90 percent of this growth deriving from yield increases and less than 10 percent from area expansion. More recently, the dissemination of Integrated Pest Management approaches has enabled pesticide application to be cut dramatically with no loss of productivity.

Despite such remarkable progress, it is a mistake to place too much optimism in technological change. New technology is often adopted in order to improve labor productivity, which in turn can raise material standards of living, but without adequate attention to the environmental impact. The impact of a particular technology depends on the nature of the technology, the size of the population deploying it, and the population's level of affluence. The World Bank, along with others, is increasing investments in more sustainable technologies, such as wind and solar energy, which have limited or benign impacts on the relations of humanity to the ecosystem that supports us all.

But the level of affluence currently enjoyed by the citizens of the OECD countries cannot be generalized to the rest of the world's current population, much less the massively larger population of the developing countries forty years from now, no matter what the improvements in technology are likely to be.

Experts in development are well aware that bringing the low income countries up to the affluence levels in OECD countries, in forty or even one hundred years, is an unrealistic goal. They may well accuse us of attacking a straw man—who ever claimed that global equality at current OECD levels was possible? We acknowledge the force of that objection, but would suggest that most politicians and most citizens have not yet accepted the unrealistic nature of this goal. Most people would accept that it is desirable for low income countries to be as rich as the North—and then leap to the false conclusion that it must therefore be possible! They are encouraged in this non sequitur by the realization that if greater equality cannot be attained by growth alone, then sharing and population control will be necessary. Politicians find it easier to revert to wishful thinking than to face those two issues. Once we wake up to reality, however, there is no further reason for dwelling on the impossible, and every reason to focus on what is possible. One can make a persuasive case that achieving per capita income levels in low income countries of $1,500 to $2,000 (rather than $21,000) is quite possible (see Serageldin, 1993b, pp. 141–143). Moreover, that level of income may provide 80 percent of the basic welfare provided by a $20,000 income, as measured by life expectancy, nutrition, education, and other measures of social welfare. But to accomplish the possible, we must stop idolizing the impossible.

The contribution to approaching global environmental sustainability differs markedly in three geographic regions. OECD's main contribution to environmental sustainability should surely be to cease its long history of environmental damage from overconsumption and pollution (corollaries to affluence under today's technology), such as greenhouse warming and ozone shield damage. The contribution of low income countries lies in stabilizing the human population. The former centrally planned economies' contribution seems to be more in accelerating the modernization of their technology, reducing acid rain by removing subsidies on dirty coal, stopping the poisoning of the land, reversing forest death, and eliminating nuclear risks. It is in OECD's self-interest to accelerate technology transfer to the former centrally planned economies and to the low income countries. It is possible that with current types of technology and production systems, the global economy has already exceeded the sustainable limits of the global ecosystem, and, therefore, manifold expansion of anything remotely resembling the present global economy would speed us from today's long run unsustainability to imminent collapse. We believe that in conflicts between political feasibility and biophysical realities, the former must eventually give way to the latter, although we cannot specify exactly how long "eventually" will be.

A DYNAMIC FORMULATION FROM SUSTAINABILITY TO SUSTAINABLE DEVELOPMENT

The foregoing discussion of Population, Affluence, and Technology is based on the static formulation of the Ehrlich and Holdren identity, presented some 20 years ago. More recent work brings nuance and shading to this generalization. Particularly important are the interrelationships among the three factors, and their links with shifts in the structure of the economy. An alternative identity, which brings out some of these relationships, and shows feedback loops, is presented in Figure 7.5 (adapted from Steer and Thomas, forthcoming).

Three questions need to be asked as we seek to monitor progress towards sustainability:

- First, given the political unreality of a voluntary decline in the overall affluence of industrial countries, how is the "pattern" of this affluence shifting? Specifically, is the economic structure of the economy shifting away from environmentally damaging activities (e.g. heavy and toxic industries) and towards less "natural capital-depleting" sectors (e.g. services). Recent research shows that structural shifts can have powerful impacts on natural resource consumption.
- Second, what is the trend in the consumption of natural resources per unit of output? Two mechanisms need to be monitored here: improvements in economic efficiency (inputs per

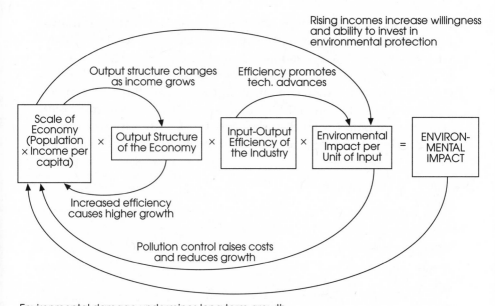

Figure 7.5 Economic activity and the environment

unit of output); and the degree of substitution away from environmentally critical inputs. Policy instruments, including taxes and user charges, can help promote such transitions, especially when the environmental costs are not captured in the marketplace.

- Third, to what extent is the pollution impact per unit of economic activity declining? Here it is important to distinguish between the innovation of new technologies, and their dissemination and application. Many of the most profound forms of environmental damage in today's world (soil erosion, lack of clean water, municipal waste, etc.) do not require new technologies, but simply the application of existing ones. This in turn requires (a) that decision-makers are persuaded that the benefits of using such technologies exceed the costs; and (b) that resources are available for putting them in place. Public policies can be targeted towards meeting both conditions.

Interactions among the driving factors: scale, structure, efficiency, technology, and investment in environmental protection, together with the key feedback loops between economic activity and human behavior, (such as the powerful impact of income on fertility) explain why in some situations economic growth and technological progress will cause increased environmental damage and in others, will do less damage. The now-famous "humps" in the relationship between certain forms of pollution and income (Figure 7.6) illustrate this point. For effective policymaking, it is essential that these various paths be disentangled so that policies may be targeted to induce changed be-

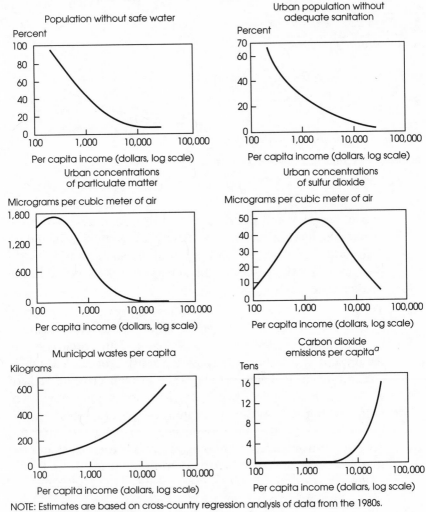

NOTE: Estimates are based on cross-country regression analysis of data from the 1980s.
[a]Emissions are from fossil fuels.
Source: Development and the Environment, World Development Report 1992.

Figure 7.6 Environmental indicators at different country income levels

havior, away from environmentally damaging inequitable growth, and towards accelerated sustainable poverty reduction.

WORLD BANK PROGRESS TOWARD SUSTAINABLE DEVELOPMENT

The World Bank's latest position on sustainable development can be found in Serageldin (1993a), so it will not be repeated here. To summarize: An entire Vice Presidency for Environmentally Sustainable Development (ESD) was created in January 1993, and a substantial staff (c. 200) is now focusing on sustainable development. ESD has started to integrate the viewpoints of the three relevant disciplines: economics, ecology, and sociology against which any technical/engineering proposals should be evaluated. ESD focuses on improved valuation of environmental concerns, building sustainability into national accounts, and how to value the future. This is amplified by Serageldin (1993a).

But beyond these conceptual concerns, the World Bank is actively pursuing a four-pronged agenda to promote ESD worldwide, through all its activities. The agenda comprises:

1. Assisting borrowing countries in promoting environmental stewardship.
2. Assessing and mitigating whatever adverse impacts are associated with Bank-financed projects.
3. Building on the positive synergies between development and the environment (often called "win-win" strategies).
4. Addressing the global environmental challenges.

PROMOTING ENVIRONMENTAL STEWARDSHIP

When it comes to assisting countries in environmental stewardship, the Bank is helping to define strategies and provide funds for environmental management as well as for projects. This last year, the Bank committed $173 million in support of environmental management, bringing up the portfolio of environmental management projects and project components financed so far up to about $500 million. This compares with a total portfolio of environmental loans amounting to about $5 billion, of which about $2 billion were committed in this last year alone. Some more details are provided in Table 7.1. In addition, the Bank is trying to help expand and disseminate knowledge by promoting the sharing of experiences between decisionmakers in the various countries.

But sound environmental stewardship is rooted in sound development and environmental strategies, which must be based on properly identifying the right priorities. These are very much country-specific.

The key point is that environmental priorities will vary from country to country, and the Bank should stand ready to assist each country with the design and implementation of its own environmentally sustainable development strategy. Each will have to address particular problems. Air pollution may be the prevalent issue in Mexico City (where the Bank is helping with a $280 million loan), but there are other forms of pollution that could be a major priority in some other cities of the developing world. Toxic wastes are the most urgent problem in parts of the former Soviet Union. In Nigeria, it could very well be the problem of overgrazing. But whatever it is, the formulation of these national strategies, we believe, should be the result of a consultative participatory process in the countries themselves. That is how we hope that the National Environmental Action Plans (NEAPs) which are now being promoted in many countries will be done.

ASSESSING AND MITIGATING ADVERSE IMPACTS

The second part of the four-point agenda is assessing and mitigating unavoidable adverse impacts of projects that the Bank agrees to finance. This requires subjecting every proposal to a rigorous environmental assessment, as well as to the traditional technical and economic assessments. Furthermore, the bank is now trying to introduce social assessment as well. The Bank has published much on environmental assessment procedures (e.g., World Book, 1992), so we will not go into detail on this point here.

BUILDING ON "WIN–WIN" STRATEGIES

Conversely, we should dwell a bit more on the third part of the four-pronged agenda: the question of building synergies between devel-

---
Table 7.1

WORLD BANK LENDING PORTFOLIO FOR ENVIRONMENTAL MANAGEMENT AND POLLUTION CONTROL APPROVED IN FISCAL YEARS 1989–1993

Country	Project	Loan/credit amount ($ millions)
New Commitments, approved in fiscal 1993		
Brazil	Water Quality and Pollution Control, Sao Paulo/Parana	245.0
Brazil	Minas Gerais Water Quality and Pollution Control	145.0
China	Southern Jiangsu Environmental Protection	250.0
India	Renewable Resources Development	190.0
Korea, Rep. of	Kqangju and Seoul Sewerage	110.0
Mexico	Transport Air Quality Management	220.0
Turkey	Bursa Water and Sanitation	129.5
Total		1,289.5
Projects under implementation, approved in fiscal 1989-1992		
Angola	Lobito-Benguela Urban Environmental Rehabilitation (92)	46.0
Brazil	National Industrial Pollution Control	50.0
Chile	Second Valparaiso Water Supply and Sewerage (91)	50.0
China	Ship Waste Disposal (92)	15.0
China	Beijing Environmental (92)	125.0
China	Tianjin Urban Development and Environment (92)	100.0
Cote d'Ivoire	Abidjan Lagoon Environment Protection (90)	21.9
Czech and Slovak Federal Republics	Power and Environmental Improvement (92)	246.0
India	Industrial Pollution Control (91)	155.6
Korea, Rep. of	Pusan and Taejon Sewerage (92)	40.0
Poland	Energy Resources Development (90)	250.0
Poland	Heat Supply Restructuring and Conservation (91)	340.0
Total		1,439.5
Portfolio total		2,729.0

Source: World Bank, 1993a.

opment and the environment. The key here is that by adopting the conceptual framework of environmental sustainability, proper development helps environmental protection, and vice versa. That is the so called "win-win" strategy. It is, to our mind, the most promising area to focus on. There are two parts: investing in people, and promoting the efficient use of resources.

Investing in people is particularly important. It is the poor who suffer the most from environmental degradation, especially women. When drought hits, it is the poor who suffer. Women are responsible for getting water, just as they have to gather fuel wood from farther and father afield all the time, and naturally they also care for the children. The solutions to better natural resource management as well as to lowering fertility all involve empowering woman. That means that investing in people, in human resource development, must pay special attention to girls' education. This is probably the single most important measure that we can adopt both for development and for the promotion of sound environmental policy over time.

Investment in people must also include population programs to recognize the pressure that the global population is putting on all of us, and these must be accomplished by the provisions of maternal and infant health care.

The efficient management of resources is the second leg of the win-win strategy. Just how inefficient the current management of resources actually is can be quite striking. Sadly, a large part of this mismanagement is currently induced by government policy. Energy subsidies in the developing world account for $230 billion a year. That is about four to five times the total volume of Official Development Assistance (ODA) going from the North to the South. That is environmentally unsound, economically unsound, and wasteful of resources that could be put to other uses.

Likewise, many of the subsidies that exist today are, in fact, for extractive and destructive industries. In the case of logging, for example, average stumpage fees are a fraction of the cost of reforestation. Among African countries sampled in 1988, the best was less than a quarter of the cost of reforestation, while the worst was running at about one percent of the cost. So subsidies were going to private loggers whereas, in fact, the full restitution to the public commons was not taking place.

If we have focused on the problems of developing countries, that is because this is where the World Bank lends. Nevertheless, the Bank is equally concerned about the inequitable use of resources worldwide and the wasteful and destructive practices being pursued by northern consumption and pollution patterns, and is trying through reports and discussions to help the requisite awareness in the North.

ADDRESSING GLOBAL CHALLENGES

The fourth part of our agenda is addressing global challenges. Here, we include national activities that have global payoffs. These are areas where much can be done to promote the global agenda from a national sovereign decision-making framework. There are, of course, global issues where the costs are local and the benefits are global. For these activities, special instruments like the Global Environmental Facility (GEF) have a crucial role to play. The GEF has also been designated as the interim funding mechanism for the two conventions on Biodiversity and Climate Change.

At the same time it is working on these global population challenges, the World Bank is also concerned with consumption in the north, and how to address the disparities between the North and the South. It is important to remind ourselves—as the UNDP's Human Development Report did so eloquently in the now famous "champagne glass" graph—that the richest 20 percent of the world receive about 83 percent of the the world's income. The poorest 20 percent of the world receive 1.4 precent. And that disparity means a huge disparity, both in terms of consumption patterns and in terms of pollution. This argues for sound strategies for people in the South because they have so few degrees of freedom, but it also certainly argues for looking again at the consumption patterns in the North. And by that, we do not mean going back to the horse-and-buggy days. Switzerland, which by no stretch of the imagination is a deprived country, has a water consumption per capita that is about one-fifth that of the United States. On energy consumption levels, the difference between Switzerland (or Japan for that matter) and the United States is also about one-half. The per capita consumption of energy in India or China is still a very small fraction of that in Switzerland or Japan. So the per capita consumption issues have to be looked at, and these argue for changes in the northern patterns, as much as they argue for sound practices in the South.

The same is true in terms of the global commons, and the contribution on the debit side, in terms of *pollution* and the use of the environment as a "sink." The contribution in terms of CO_2 emissions, or in terms of global waste production and pollution, show the same types of disparities. They are also very large. India's per capita contribution of average annual tons of carbon emitted into the atmosphere is very small compared to Canada or the United States, and this is true of most developing countries, except for the former USSR, where levels are relatively high because of the nature of their industrial activities.

Such disparities encourage one to think in terms of tradeable permits. Low-income countries with a large population could trade permits based on proportional population rights to use environmental services (both to consume and to pollute) with some of the richer countries. While this is not currently on the agenda of international negotiations, there is something there for all of us to reflect on.

In addition to the agenda alluded to above, the Bank has substantially increased direct environmental investments (Table 7.1): a record $2 billion for 23 projects to assist developing countries in improved environmental management. This represents a doubling over one year ago, and a 35-fold increase over lending five years ago. Financing for the enabling conditions of sustainability also soared: $180 million for population, $2 billion for education, $5 billion for poverty alleviation. In addition, more than 30 countries have prepared national environmental action plans with assistance from the Bank. The Bank is convinced that environmental sustainability is essential and costs less than unsustainable development.

A major obstacle to promoting policies that foster sustainability to date has been the incomplete measurement of income and investment, particularly the failure to reflect the use or deterioration of natural capital (Steer and Lutz, 1993). To correct this failure, the Bank is promoting improvements in National Income Accounts (SNA) (Ahmad et al., 1989; Lutz et al., 1993; El Serafy, 1993). Environmentally adjusted SNA has massive policy implications for most developing countries.[4] Without environmentally adjusted SNA for example, we cannot judge if an economy is genuinely growing or merely living unsustain-

---

[4] Economics Nobel Laureate Robert Solow has recently retreated from his 1973 position that natural capital is unimportant: "The world can, in effect, get along well without natural resources"(Solow, 1974. p. 11). In 1992, he concluded that the U.S. GDP may change only 1–2% if environmentally adjusted (Solow, 1992). Perhaps this seems a small concession, but during the discussion period of his 1992 RFF lecture, he recognized that the developing countries rely to a vastly greater extent on natural capital than does the United States.

ably on asset liquidation beyond its true income; whether the balance of payments is in surplus or deficit on current account, or whether the exchange rate needs to be changed.

CONCLUSION

This paper represents our current views on the concept of environmental sustainability. This is ongoing work, part of which is being done in the World Bank, much of it being done by concerned scholars around the world. Our aim has been to make a modest contribution to the debate on the essence of sustainability. We fully expect the concept to be refined in the coming months and years. By its actions, it is clear that the World Bank is taking environmental sustainability seriously indeed. But this conceptualization is far from an academic exercise. The monumental challenge of ensuring that possibly 10 billion people are decently fed and housed within less than two human generations—without damaging the environment on which we all depend—means that the goal of environmental sustainability must be reached as soon as humanly possible.

CHAPTER 8

Vision of the South

In the five decades since the end of World War II, the world has witnessed an impressive record of economic growth in the industrialized countries, transforming those societies into modern ones with high standards of living. This same process of economic growth has, however, two important deficiencies: first, it has degraded the global environment; second, it has created benefits of prosperity and progress which largely bypassed the Southern hemisphere leaving those developing countries in a condition of poverty and underdevelopment which also degrades the environment. Such a pattern of development is clearly not sustainable.

Four-and-a-half billion people, or 85 percent of world's total population, occupying 72 percent of the land area of the globe, belong in the category of low- and middle-income countries of the South, with a per capita income ranging from US$80 (Mozambique) to US$7,820 (Saudi Arabia) in 1991 (International Bank for Reconstruction and Development, 1993b, pp. 238–239). If these countries, in their endeavor to raise per capita income, are following the same path as the industrialized countries, the impact will be catastrophic to the global environment. It is therefore necessary for the South to follow a different path of development: one that makes possible the eradication of poverty and at the same time does not degrade the environment. This calls for an elaborated pattern of sustainable development that still meets the interests of the South.

Income per capita varies widely among the developing countries, which reveals the differences in population; available natural resource endowment; the level of economic activities; the countries' respective social, economic, and political structure; etc. In general however, common characteristics can be identified which distinguish the South from the North. There are many characteristics of underdevelopment that can be identified in the South. The first and most striking characteristic is poverty (as indicated by low income per capita combined with unequal distribution of income within and between countries). According to the *World Development Report* (1990) the burden of poverty is spread unevenly among the regions of the developing countries, among countries within those regions, and among localities within those countries. South Asia has a total of 46.4 percent, East Asia 25 percent, Sub-Saharan Africa 16.1 percent, Latin America and the Caribbean 16.1 percent, while Europe, the Middle East, and North Africa have 5.9 percent of the world's poor (International Bank for Reconstruction and Development, 1990, p. 2).

Development that takes place without touching the poor will widen income inequality and hence raise social instability, which in the long run makes development unsustainable.

The second characteristic of underdevelopment is the primary sector-based and unbalanced structure of the economy. Most of the developing countries are dependent in their employment, income, and foreign exchange earnings on narrow-based agricultural commodities or mineral resources. Due to lack of capital and skills, the processing of these primary products and the creation of the bulk of value added is accomplished outside the boundaries of the producing country. In the world market, the prices of raw materials suffer a declining trend, and hence reduce their contribution to the income of the country.

The third characteristic is the structural deficit in the balance of payment. Most of the developing countries import more than they export. Practically all capital equipment; most of the production materials; and, for many developing countries also, most food is imported. The resulting balance of payment deficit is covered by foreign debt. This raises the dependency of the South's economies on the North.

The fourth characteristic is macroeconomic instability, as revealed by a high rate of inflation. Inflation can destabilize the economy and squander natural resources, which is not conducive to sustainable development.

And last but very dominant is the population factor. It is expected that the rate of population growth will still be significant in spite of an active family planning program. The population of the South is expected to increase from 4.5 billion, or 85 percent, of the world's total population in 1991 to 7.3 billion people or 89 percent of the world's total population in 2025. A large part of the population in the South suffers a high prevalence of malnutrition (especially for those under five years of age), low life expectancy at birth, and a low level of formal education. While quantitatively the size of the population is large, qualitatively its productive capacity is low. Because of improved health science and technology, ways and means are now available that reduce crude death rate faster than crude birth rate. This acts more as a liability than an asset for sustainable development.

POVERTY

Sustainable development as pursued by the South must first and foremost focus its efforts on meeting the developmental challenges which are the causes of environmental degradation. In this connection, the first objective of sustainable development must be to eradicate poverty. Poverty, as defined by the *World Development Report 1990*, refers to the inability to attain a minimal standard of living. This inability is caused not by external factors but internal ones such as: no skill, low education, poor health, and low capacity to respond to income-earning opportunities. On the other hand it is also caused by factors external to the poor, such as: no access to land tenure, minerals, or other natural resources; no access to credits or other financial resources; no access to markets; no access to technology; and no access to productive infrastructure like irrigated water, electricity, transportation, etc.

To overcome this inability, development should be geared to ensure the poor the provision of basic needs such as nutritious food, clothes, shelter, basic education, health facilities, and clean water. All of this requires a development policy that is focused on engaging the poor in productive employment.

PRIMARY SECTOR-BASED AND UNBALANCED STRUCTURE OF THE ECONOMY

Productive employment can be created by diversifying the primary sector-based economy into a secondary and tertiary sector-based economy. The experiences of the six economies of Indonesia, Japan, Korea, Malaysia, Thailand, Taiwan, and China have shown that average annual *growth* rate of agricultural income has drastically increased during 1965–1988, even though agriculture's share of output and employment has declined sharply during the same period. The various factors that contributed to the success of agriculture in these economies are land reform, agricultural extension services, reasonably good infrastructure, and heavy investments in rural areas (International Bank for Reconstruction and Development, 1993a, p. 32). High productivity in agriculture strengthened the competitive position of these economies, and allowed them to raise their market share in the commodity market in spite of declining commodity prices. The share of agriculture in gross domestic output did not decline because of reduced output, but because of the increased share of other non-agricultural output. It is important to note that it is possible to move from an unbalanced primary sector based economy toward a self-reliant growth economy with diversification and higher value added.

This process of economic development, which is currently considered successful in some Asian countries, has shaped the pattern for the desired economic state of the developing world. The phrase "developing country" suggests a state of progress, however, which is probably not unfolding at all, and that is counterintuitive to our mental mindset. The flow of aid money to the South, and the expansionist behavior of some new economies all create impressions of development.

SOUTH-TO-NORTH RESOURCE FLOWS

It is believed that the aid given to the South by the North is substantial, and that in the end the South will benefit. However, in reality, and caused by the workings of unbalanced international economic structures, there is a massive outflow of financial and economic resources from the South, with the North countries as beneficiaries. Issues such as external debt, commodity prices, terms of trade, investments, and terms of technology transfer need consideration; but prior to, and after the UNCED, the North had made it clear that reform of international structures and relations are out of the question. The northern UNCED commitment to increase aid, to provide new and additional funds, and to reach the 0.7 percent target, yet without specifying a target year, is an important step forward. This will not significantly or adequately compensate for the ongoing massive loss of resources that the South experiences due to the structure of the world economy, and which seriously impedes the possibility for many Southern countries to implement programs for the transition to sustainable development.

If we review the various mechanisms by which resources are flowing out from South countries, we will then have to make efforts to reduce, stem, and (if at all possible) reverse the outflows. Reforms to international structures and to global economic relations will be required if this is to be done. The facts and figures follow.

In international trade, a majority of Southern countries are still exporters of mainly raw materials and commodities (primarily to the North) and importers of mainly industrial products (mostly from the North). The level of commodity prices in general has been falling when compared to the level of prices of manufactured goods, and often even in nominal terms. As a result, commodities have suffered adverse terms of trade vis-a-vis manufactured goods, with the South experiencing very significant losses. These losses on account of adverse terms of trade deplete the South of foreign exchange as well as real income, as the same unit of export earns in exchange a lower and lower quantity of imports.

According to U.N. Secretariat data, the US dollar combined price index of nonfuel primary commodities exported by developing countries has fallen from 171 in 1980 to 100 in the index's base year 1985 to 115 in 1992 (United Nations, 1991, p. 224; United Nations, 1993, p. 227). In the same three years, the price index of manufactures exported by developed market economies had moved from 116 to 100 and then risen to 164. As a result of these movements (the fall in commodity prices and the rise in prices of manufactures), the terms of trade of commodities vis-a-vis manufactures (or the real prices of commodities) had declined very substantially from 147 in 1980 to 100 in 1985 to 80 in 1990 and even further to 71 in 1992.

These figures are so alarming that it would be difficult to exaggerate the sharpness of the decline and the seriousness of the losses for the South. Between 1980 and 1992, nonfuel commodity prices had fallen by 52 percent on average . In other words, on average, whilst in 1980, 100 units of a Southern country commodity export could buy 100 units of manufactured product imported from the North, the same 100 units of commodity export could only pay for 48 units of the same imported manufactured product in 1992. There is thus a loss of real income amounting to 52 units of imports for every 100 units of exports, so a great loss of real economic resources for the South.

The data above refer to nonfuel commodity prices, as in the past the changes in oil prices have had a big effect on movements in commodity prices overall. The indices on nonfuel commodity prices better reflect the position of a majority of Southern countries. However, oil exporting developing countries (which in the 1970's enjoyed income gains due to dramatic oil price increases) themselves suffered massive losses in the 1980's due to the decline in oil prices. The index for the price of crude petroleum fell from 123 in 1981 to 100 in 1985 to 63 in 1992, a drop of 49 percent between 1981 and 1992 (United Nations, 1992, p. 206; United Nations, 1993, p. 227). The real price of crude oil (after deflating by the price index for manufactures) fell even much more sharply: from 113 (in 1981) to 100 (1985) to 38 (1992). The real price of crude oil in 1992 was only one-third of its level in 1981. If we add the loss suffered by oil exporters, the situation of the South would be even much worse.

The effects of this terms-of-trade decline on income in the South have been devastating. Given the importance of this issue, it is surprising that there have been so few published materials on estimates of income losses to the South on account of terms of trade losses. One of the few estimates was made by the UNCED secretariat: "The impact of these price movements on country groupings and individual countries varied according to the composition of their exports. Commodity-dependent countries lost ground, while countries with more diversified exports and a high share of manufactured exports faced neutral or, in some cases, improving terms of trade. But the fact is that, on the whole, developing countries were adversely affected by changes in terms of trade. Terms-of-trade losses of one country are the gain of another. Industrial countries benefited from the changes in terms of trade in the 1980's" (UNCED, 1991, p. 8).

The UNCED paper examines the impact of unfavourable terms of trade on two groups of countries: Sub-Saharan Africa and middle-income highly-indebted countries. The results are astonishing. For Sub-Saharan Africa, the terms of trade fell from 100 in 1980 to 72 in 1989. This 28 percent fall in terms of trade led to a loss of income of US$16 billion in 1989 alone, equivalent to 9.1 percent of the Sub-Saharan Africa's combined GDP. In the four years 1986–1989, Sub-Saharan Africa suffered a total $55.9 billion of income losses due to terms-of-trade decline, equivalent to 15–16 per cent of GDP in 1987 (UNCED, 1991, p. 9).

For 15 middle-income highly indebted countries in the study (Argentina, Bolivia, Brazil, Chile, Colombia, Cote d'Ivoire, Ecuador, Mexico, Morocco, Nigeria, Peru, Philippines, Uruguay, Venezuela, and Yugoslavia), the losses were also very high. Their combined terms of trade also fell by 28 percent from 100 in 1980 to 72 in 1989. This caused an income loss of $45 billion in 1989 alone, amounting to 5.6 percent of their combined GDP. In 1981–1989, the total income losses due to terms-of-trade decline were $247.3 billion. The losses in 1986–1989 were, on average, $45 billion per year, or 5–6 per cent of GDP (UNCED, 1991, p. 10).

The UNCED paper also calculates net financial transfers from the 15 highly indebted countries, arising from investments, debt repayment, and aid flows. In 1989 there was a net transfer from these countries of $35 billion, equivalent to 4.4 percent of GDP. When added to the income loss from terms of trade of $45.1 billion in the same year, the total loss was $80.1 billion, or 10 percent of the GDP. Thus these countries lost 10 percent of their total domestic resource availability on account of these factors alone.

The combined losses of the two groups of countries (Sub Saharan Africa and the 15 middle-income indebted countries) due to terms of trade decline were $61 billion in 1989. The total income losses due to trade terms for the South as a whole would have been so much larger than even this, as most countries (which also suffered losses) are not counted in the $61 billion figure, and moreover the loss would have been higher if a base year earlier than 1980 were to be taken (since real commodity prices have been falling since the 1950s).

In 1972, the non-oil developing countries were losing US$10 billion (over 20 percent of the value of their aggregate exports) due to the lower level of their terms of trade compared to the mid-1950s

(Clairmonte, 1975, p. 1186). In the 1970s the terms of trade fell further. According to UNCTAD estimates, Third World commodity export earnings in 1978 were US$45 billion lower than what they would have been if prices had remained at their 1969 levels. At that time, UNCTAD predicted that prices would fall further in the 1980s (a prediction that came true), and it had estimated that by 1990 the shortfall could reach $186 billion (Khor Kok Peng, 1983, pp. 21–23). Augustin Papic, a former South Commission member, has written that "invisible transfers" from the South due to losses from terms of trade total close to $200 billion a year, according to some estimates (Papic, 1991, p. 17).

The income losses from falling terms of trade probably constitute the largest single mechanism by which real economic resources are transferred from South to North. They severely constrain the resources available for sustainable development, and moreover compound the debt problem for many countries as the loss of foreign exchange makes it extremely difficult for many countries, and impossible for some, to meet their debt servicing obligations. Given the importance of this subject, it is surprising that too little attention had been given to the income losses from trade terms, and this should be rectified.

The external debt crisis, which originated in the 1970s, has been another major source of financial drain on the South. The inability of many Southern countries to meet their debt obligations is due to many factors, including such domestic factors as the proliferation of economically unfeasible and socially inappropriate projects and financial mismanagement.

However, the problem was also largely induced by international factors often beyond the control of the indebted countries, including the deterioration in commodity export prices and the devastating effects this had (and is having) on income and foreign exchange, the increase in interest rates in the 1980s, and for some countries, changes in the relative exchange rates of the currencies in which the debts are denominated.

The drain on resources from the South due to debt servicing has been enormous. Due to the rapid accumulation of foreign debts, Southern countries increased their annual interest payments tenfold from US$2.5 billion in 1970 to $25.7 billion in 1979 and within two years they doubled again to $51 billion in 1981. In the period 1970 to 1982, a total of $239 billion of interest was paid out by the Southern countries on their external debt (Khor Kok Peng, 1983, p. 26).

The situation worsened greatly in the 1980s. According to U.N. Secretariat data, interest payments of capital-importing developing countries were $46 billion in 1980 and $60–70 billion annually in 1981–1992 (United Nations, 1991, p. 237; United Nations, 1993, p. 253). Total interest payments in the period 1980–1992 were $771.3 billion.

Interest payment is only part of overall debt servicing, which also includes repayment of principal. Debt servicing of capital importing developing countries rose from $90 billion in 1980 to $117 billion in 1984, $141 billion in 1989 and $158 billion in 1992. In 1980–1992, total debt-servicing flows of capital-importing developing countries were $1,662.2 billion. This comprised $771.3 billion in interest payment and $890.9 billion in principal repayment.

Despite these huge debt-servicing flows, the countries concerned were even more indebted at the end of the period. Their total external debt rose from $567 billion in 1980 to $1,066 billion in 1986 and $1,419 billion in 1992 (United Nations, 1991, p. 236; United Nations, 1993, p. 249). The debt stock rose by $852 billion between 1980 and 1992. Moreover, at the end of 1991 the countries had principal arrears on long-term debt of $49.8 billion and interest arrears on long-term debt of $35.4 billion.

Between 1980 and 1989, principal repayments were $891 billion while there was a net increase in debt of $852 billion. Thus, gross new disbursements of loans in that period were $1,743 billion. A very large part of this was used to repay old loans and meet interest payments.

The situation then was this: in 1980 the capital-importing developing countries had $567 billion in debt. In the 12 years to 1992 they borrowed another $1,743 billion, but much of this was used to meet debt servicing obligations as debt-service flows totalled $1,662 billion. As a result, by the end of the period, the countries had an even much larger debt stock of $1,419 billion, which will require even larger debt-servicing outflows in future.

The situation can also be put another way. In 1980 these countries had $567 billion in debt. Over the next 12 years they repaid $891 billion in principal and also paid out $771 billion in interest. But since much of this debt servicing had to be financed by new loans, they were saddled at the end of 1992 with a much larger debt stock of $1,419 billion, two-and-a-half times larger than in 1980.

The inflow of foreign investment is usually seen as a source of capital transfer from North to South. However, it can also be a major source of outflow of resources from the South. This is because while foreign direct investment may enter a Southern country through one door (the capital account in the balance of payments), it can also lead to an exit through another door due to repatriation of profits and dividends (outflow of investment income in the current account of the balance of payments).

In recent years there has been an increased inflow of foreign direct investment to the South. The net foreign direct investment flow to capital-importing developing countries was $9.5 billion in 1987, and rose to $23 billion in 1991 and $25 billion in 1992 (United Nations, 1993, p. 238). However, most of this went to a relatively few countries.

As the foreign firms establish themselves, they eventually derive higher profits and dividends, a large part of which may be repatriated to the headquarter companies. Profit outflow may eventually be more than new inflow of foreign investment, which can then result in a net outflow of resources. From 1987 to 1992, there has been an increase in new foreign investments to the South, which have exceeded the outflow of profits, but in 1980-1986, the net profits of foreign companies exceeded new net investment flows. In 1980-1983, net profit outflow from the South averaged $9-10 billion annually, and in 1989-1992 it averaged $10-11 billion. For the period 1980-1992 total net outflow of dividends and profits from capital-importing developing countries was $122 billion. In 1980-1986, profit outflow exceeded new investments by $15.5 billion; in 1987-1992, investments exceeded profits by $48.3 billion (United Nations, 1990, p. 82; United Nations, 1993, p. 238).

These statistics on profits may be significantly understating the outflows because foreign company profits are often transferred out in hidden ways, such as through "transfer pricing," in order to reduce tax or, in the case of local-foreign joint ventures, to reduce the profit share accruing to the local partner and thus maximise the foreign multinational's overall profits.

The extent of transfer pricing and the volume of profits derived from foreign firms due to this practice is difficult to quantify, and can only be gauged by the few studies that have been made. August Papic (1991, p. 11) quotes UNCTAD studies which estimated that pharmaceutical transnationals charged their Latin American subsidiary companies between 33 and 314 percent above world prices for the inputs they provided; and in the electrical goods industry, the markup was estimated to be between 24 and 81 percent.

An earlier study in Colombia in 1968 found that Colombian subsidiaries of pharmaceutical multinationals were overcharged 155 percent (compared to international prices quoted in Europe and the US), and the average profit rate of foreign pharmaceutical firms was actually 79 percent and not the 6 percent reported (Khor Kok Peng, 1983, p. 28). Due to the transfer price system, a large part of foreign firms' real profits is not revealed in the figures for repatriation of dividends but are hidden in the inflated import values and deflated export values.

Another mechanism by which foreign investment leads to resource outflows from the South is the increasing participation of Northern-owned funds in the stock markets of Southern countries. Termed "emerging markets," many of the Southern countries' equity markets have been providing higher rates of return than their counterparts in the North. There has thus been a large increase in foreign portfolio investment in these Southern countries, and the foreign investors have been making large profits from capital gains.

Since the foreign funds are usually in large blocks, and operated by experienced investment agencies, they are more likely than small local players to make gains from the fluctuations of the stock markets, and may also contribute to the speculative aspects of the markets. Often, therefore, the foreign fund players make their profits at the expense of local people. When these gains are repatriated, they contribute to the South–North transfer of resources.

There is also a large outflow of resources from the South due to payment for access and use of technology owned by Northern corporations. These "technical payments" include royalties for patent use, license fees, and payments for process know-how and trademarks. Also, the foreign subsidiary or branch operating in the South usually pays a "management fee" or "technical fee" to the parent company in the developed country. Technology contracts are signed between companies of the same multinational based in the developed and developing countries, specifying the amount of royalty or fees to be paid to the parent company or technology supplier, such payments usually specified as a percentage of sales revenue.

The significance of these technical payments in terms of what they reveal about the real level of foreign profits can be gauged from a study of multinational companies in six developing countries by Sanjay Lall and Paul Streeten, who found that these technical payments were an important vehicle for remitting profits from the subsidiaries to the parent companies. The technical payments constituted 0.5 to 2.5 per cent of sales, with an average over the six countries of 1.3 percent. As equivalents to the declared post-tax remitted dividends of the firms, the technical payments were 12-16 percent in Kenya and Jamaica; 16 percent for foreign-majority owned firms and 135 percent for foreign-minority owned firms in India; 420 percent for foreign-majority and 60 percent for foreign-minority owned firms in Iran; 66 percent for foreign-controlled firms in Colombia; and 52 percent for foreign-controlled firms in Malaysia (Khor Kok Peng, 1983, p. 28).

An earlier study by UNCTAD (1981, p. 37), examining both technical payments and transfer pricing together, shows the vast transfer of resources from South to North on account of its technological dependence. The 1981 report states: "Dependence on foreign technology carries a high price. Direct payments by developing countries for the use of patents, licences, trade marks, process know-how, and technical services amounted to US$1.8 billion in 1968, and probably now stand at US$9-10 billion. The direct costs, however, constitute only a part of the total costs. The indirect costs resulting from overpricing imports of intermediate products and equipment, profits on capitalization of know-how and price markups are much heavier. In addition, there are other real costs, or benefits forgone, as a result of delayed or inadequate transfers, the transfer of inappropriate technology and, above all, the non-transfer of technology. It is estimated that the total cost of technological dependence may be as high as US$30-50 billion a year."

Since the UNCTAD estimate of the early 1980s, the outflows from the South on account of transfer pricing and technical payments must have increased far beyond $30-50 billion annually. These payments can also be expected to rise dramatically in the future following the Uruguay Round agreement on trade-related intellectual property rights (TRIPs), which will oblige Southern countries to introduce or tighten existing national laws relating to intellectual property to conform to Northern legal standards. Northern-owned transnational companies will be the main beneficiaries of TRIPs, as the payments of royalties, license fees and other technical payments can be expected to increase rapidly after the passage of new IPR regimes in the South. Indeed, technical payments and hidden profits in transfer pricing will soon compete with terms-of-trade losses and debt servicing as the major sources of resource outflow from the South.

Another form of resource transfer from South to North is the migration of highly skilled manpower, with the South thus contributing "intellectual technology" to the North. The brain drain is so prevalent and significant that far from the North providing technical assistance to the South, the South is, in net terms, providing much more technical aid to the North.

Through the brain drain, the South loses economic resources in at least two ways. First a substantial amount of money (usually public money) had been expended in developing countries to educate and train professional or skilled personnel in the expectation that they would in turn contribute their skills towards the nation's development; when the personnel migrate, the funds expended on their training are thus lost to the country. Secondly, the country also suffers a cost in terms of the loss of stream of potential and future economic benefits that could have arisen from the application of the skills of the migrating personnel.

The brain drain thus represents a flow of economic capital from South to North. In 1970, there were 11,236 skilled immigrants entering the US alone, representing a brain drain or reverse transfer of technology amounting to $3.7 billion in that year alone (Clairmonte, 1975, p. 1187).

A 1982 UNCTAD study estimates that the industrial countries gained about $51 billion of "human capital" from 1961 to 1972 as a result of migration of professionals from the Third World (Khor Kok Peng, 1983, p. 29). It also reveals that the US, Britain, and Canada gave $46 billion in development aid to poor countries between 1961 and 1972, but received $51 billion of "human capital" due to the brain drain during the same period.

Canada gained seven times more in migrant talent than it spent on educational aid, Britain gained 2.8 times as much, while the gain to the US was 86 percent of the aid Washington gave out. As the study concluded: "The migration of manpower from developing to developed countries is not just a movement of persons. It is a real transfer of productive resources from poor to rich countries."

Capital flight, or the movement of surplus savings or investment funds of citizens and companies in the South to countries in the North is also a manifestation of South-to-North resource flows. Capital flight takes many forms, including the purchase of property such as houses, land, and hotels abroad, the placement of savings in overseas banks, and investment in stock markets abroad.

Ironically, the foreign assets of politicians, citizens, and companies of some Third World countries that are highly indebted are worth a sizeable portion of their countries' external debt. According to International Monetary Fund estimates, the annual capital flight from 13 highly indebted countries was $180 billion in 1988, much higher than the $47.3 billion in 1978 (Papic, 1981, p. 10).

Many Third World countries have suffered losses caused by changes (often sudden and sharp) in the exchange rates of major currencies. Sometimes the changes in the relative rates of exchange of these major currencies are deliberately induced by decisions of the governments of the North, for instance at Group of 7 Summit meetings. Countries in the South are greatly affected by such decisions, often adversely, but are not consulted, have no say, and perhaps the effects on them are not even a factor considered when such decisions are made.

The effects can be in different forms. A Third World country may hold its international reserves in a number of currencies or in investments in some developed countries. Should the currency of a country where a substantial portion of the reserves are held fall sharply in value, the Third World country involved would suffer a severe loss. This happened when the British sterling fell sharply in value in 1993, and some Third World countries suffered major losses. As much of the gains were made in the North (perhaps by currency traders and speculators) there was a South-to-North resource transfer.

A Third World country holding external debts denominated in a currency whose value appreciates sharply will also suffer a loss as the debt stock would have risen in terms of the local currency (although it may remain the same foreign currency), and it would take more in terms of local currency to service the debt.

Although at the UNCED it was declared officially that aid should increase, in reality, since 1992, the aid position has turned from disappointing to alarming. Instead of aid increases, many Northern governments, including some of those that traditionally had been more sympathetic, have announced reductions in their aid budgets. The Secretariat of the Commission on Sustainable Development has reported that "many bilateral donors have cut back their aid programs in response to budgetary pressures, causing a further shortfall from the target of ODA flows of 0.7 percent of GNP." It added: "The outlook for the growth of aid from industrialized countries is bleak."

According to a CSD Secretariat's updated report (February, 1993) on financial resources, aggregate aid from OECD members increased in nominal terms by 5.8 percent, to $59.9 billion in 1992, but this was a 0.3 percent decline in real terms. Bilateral aid fell 6 percent in real terms and, in particular, bilateral grants dropped 12 percent. This was offset by a 19 percent increase in the OECD members' contributions to multilateral agencies.

THE POPULATION FACTOR

Sustainable development is difficult to implement in developing countries because of the negative impact of population pressure. Between 1991, with 5.3 billion people; and 2025, with a projected population of 8.2 billion, approximately 96 percent of this increased population will be in developing countries. Facing such an upsurge of population, it is important that the South follows an active population policy which should at least contain the following points:

- first, to reduce the rate of population growth in order to reach a stable population as soon as possible. This requires not only an active family planning program but also programs "beyond family planning," such as improving the status of women, raising the welfare of families; increasing education, employment, etc.;
- second, to promote the mobility of population from high population density areas into low density areas to reduce the pressure of overexploitation of natural resources. Related to this is the need to manage the increase rate of urbanization which creates mega cities;
- third, to raise the quality of the population, either physical or nonphysical. The physical quality of population is enhanced through the improvement of nutritional intake, the reduction of morbidity rate, achieving physical fitness, etc. Nonphysical quality of the population is enhanced through training, education, and cultural and other human developmental endeavors. Both physical and nonphysical enhancement of population are geared towards raising the productivity and quality of humans.

TOWARDS SUSTAINABLE DEVELOPMENT

At the initial stage of development, up to the mid nineteen-eighties, most developing countries applied a centrally-planned market economy. Government intervention in the market economy was most significant. The results thus far have not been satisfactory.

Through the application of Structural Adjustments Policies as promoted by the World Bank, many developing countries are shifting towards a more deregulated and decontrolled market economy. This is in particular true for the High Economic Performing Asian Economies, which have shown impressive achievements in past years. The major cause of the East Asia miracle may perhaps lie in reduced inefficiency in the management of the economy.

Inefficiency is caused by market failure, referring to the failure of the market to reflect the full costs of production in the price of traded products and inputs. Inefficiency can also be caused by policy failures revealed by government intervention such as tax distortions, subsidies, quotas, interest rate ceilings, and others which distort a well-functioning market, exacerbate an existing market failure, or fail to establish the foundations for the market to function efficiently (Pearce and Warford, 1993, p. 173).

Both market and policy failures are also the causes of environmental destruction. Common resources are unpriced, waste and pollution have no markets. Environmental costs are not captured by the governments when interfering in the markets. This is still the other side of the glittering coins of East Asia: the destruction of the environment is, as expected, vast.

In order to overcome these failures and to prevent environmental destruction, it is suggested "that the most cost-effective intervention for mitigating market failures is to improve the functioning of the market by eliminating policy induced distortions, establishing secure property rights over resources, internalizing the costs of external side effects through pricing and fiscal instruments, encouraging competition, allowing the free flow of information, and reducing uncertainty through more stable and predictable policies and politics" (Panayotou, 1993, p. 30). This suggestion implies that environment is merged into economic development, which provides the base for sustainable development. In line with this development, current national income measures have to be reviewed and adjusted (Ahmad et al., 1989).

In order to move from conventional development towards sustainable development as elaborated above, the South requires substantial resources to finance Environmental Adjustment Policies and Programs designed to:

* meet the developmental challenges;
* meet the environmental challenges;
* enhance resource use management;
* enhance pollution control and waste management;
* reduce inefficiency in the management of sustainable development by correct market and policy failures.

This argues for sound strategies for people in the South because they have so few degrees of freedom, but it also certainly argues for looking again at the consumption patterns in the North. And by that, we do not mean going back to the horse-and-buggy days. Switzerland, which is by no stretch of the imagination a deprived country, has a water consumption per capita that is about one-fifth that of the United States. On energy consumption levels, the difference between Switzerland (or Japan for that matter) and the United States is also about one-half. The per capita consumption of energy in India or China is still a very small fraction of that in Switzerland or Japan. So the per capita consumption issues have to be looked at, and these argue for changes in the northern patterns, as much as they argue for sound practices in the South.

Financial resources can be mobilized by forging North-South as well as South-South cooperation through trade, investment, technology transfer, capacity building, debt relief, and development aid. The newly established World Trade Organization needs to remove the barriers that hinder the trade of increased value added products from the South to the North. The World Bank and the IMF need to reassess the economies of the North as well, to allow structural adjustments in their production and consumption patterns in order to allow free flow of investments into the South. Clean technology transfer needs to be promoted, renewable energy technology needs to be boosted. Human resource development and capacity building in sustainable development needs to be encouraged. The international financial institutions and creditor countries need to reduce outstanding debt and debt services to make these sources available for sustainable development. Finally, fresh aid can be mobilized by restructuring the tax system to shift the tax burden from environmentally friendly activities such as income from investment in sustainable production and consumption.

With this effort being implemented in the spirit of global partnership, it is possible to forge ahead with sustainable development as seen from the South. And moreover, formal accounting of aid flows is necessary: we propose a yearly World Bank/IMF report on the actual and final flows of aid and capital.

CHAPTER 9

Indicators for Measuring Welfare

There is widespread consensus that GNP is a misleading measure of welfare or well-being, as discussed in earlier chapters. Despite caveats from economists that the GNP was never intended to be used as a measure of welfare, it has come to be used as the primary macroeconomic indicator of welfare or progress in modern society. Various approaches have been pursued to find alternative indicators to remedy the most deficient and criticised shortcomings of the GNP, as has been treated in Chapter 5. Here we present the various lines of research that have been pursued in remedying the GNP's shortcomings in these contexts.

WELFARE, WELL-BEING, AND WEALTH

To fully appreciate the broadness of the task of measuring welfare, it is important to be explicit about what should be measured.[1]

- *Welfare* is a state derived from satisfaction of wants or needs evoked by our dealings with scarce means.
- *Well-being* is a state derived from the satisfaction of wants or needs evoked by our dealings with both scarce means and non-economic factors. Well-being, thus defined, therefore includes welfare but goes beyond it.

In practice, welfare and well-being are used interchangeably. Strictly speaking welfare is what economists consider the goal of economic activity. Its components include income, health, the quality and quantity of work and leisure, environmental quality, and personal and social security. Well-being, or happiness, is probably a function of welfare, but there is no automatic relationship between them. Well-being is a subjective condition: some people experience well-being in conditions that others find profoundly uncomfortable (Ekins, 1992, p. 46).

Wants are treated by neoclassical economics (the current paradigm) as the same as needs. Satisfaction of wants (via the economic system) is viewed by economists as unequivocally good—and the more satisfaction of wants the better—irrespective of scale. Ekins (1992) makes a distinction in which the lack of fulfilment of needs results in damage to an individual. Welfare, he writes, can come not only from the satisfaction of wants but their diminution. This is a concept that is scarcely treated in neoclassical economics but has important implications for sustainability and any attempts to measure welfare which use consumption or income as a basis (in particular, the GNP per capita).

Wealth is usually interpreted as the amount of money at one's disposal. Money represents a claim on produced goods and services. However, if the purpose of production is to generate welfare, and if wealth is also a means to that end, then wealth should be interpreted more broadly to include all the economic factors or conditions con-

---

[1] These definitions are based on the study of "Real Wealth" by Paul Ekins (UK), Theo Potma and Roefie Hueting (NL) and E.K. Seifert (FRG).

tributing to welfare, as expressed by the term "real wealth." Real wealth can be thought of as a means to an end, a means by which people can develop their innate potential and lead fulfilling and satisfying lives (Ekins, 1992).

Environmental quantity and quality play an important role in determining wealth and welfare. One additional definition is therefore useful regarding the role of environment: Environmental functions are possible uses, including passive ones, of our physical surroundings; that is water, air, soil, land, plant and animal species, natural resources. As functions become scarce goods they influence welfare, and consequently, well-being. As long as they are still free goods (exceptional) they affect well-being.

Figure 9.1 is a simplified illustration of the influences on well-

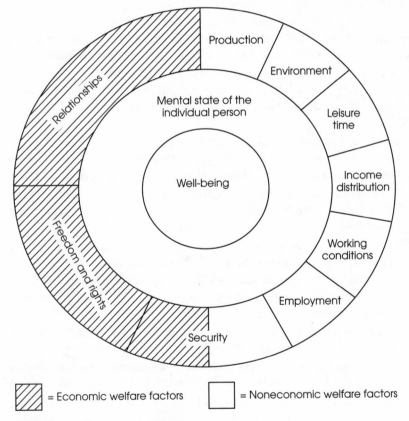

Figure 9.1   Influences on well-being

Source: "Real Wealth" by Paul Ekins, Theo Potma, Roefie Hueting, and Eberhard Seifert.

being. Each of the factors in the outer ring represent elements of economic or non-economic welfare. Measurement of real wealth must cover a wide range of aspects, including sustainable development which meets the needs of the present generation without compromising the ability of future generations to meet their own needs.

Against this background—economic- and non-economic elements of welfare—a variety of approaches have emerged. Together they constitute an evolving framework of complementary, although often disparate concepts of measuring and valuing societal progress. In attempts to improve the measurement of welfare, two main lines of alternatives have been pursued. First, efforts to supplement GDP by measuring and evaluating aspects of welfare using social and environmental indicators (basket approach). Second, attempts to produce alternative indicators, directly or indirectly derived from GDP, to measure welfare (index approach). The different strands of research are described in the following sections.

DISAGGREGATED SOCIAL INDICATORS OF QUALITY OF LIFE AND WELFARE

The social indicator movement, in reaction to the shortcomings of the GNP as a welfare indicator, began in the 1960s and was expressive of a new concern to become better informed about social goals, processes, and achievements (Ekins, 1992). Three social scientific traditions were established for measuring social progress or quality of life:[2]

- The measurement of subjective well-being, based on personal reports (derived from surveys), attempts to measure which groups of people are most satisfied with their lives. This is a subjective method commonly used in Anglo-Saxon regions.
- The measurement of living conditions, based on individuals' resources and opportunities such as income, estate, education,

---

[2] A good introduction to the "Social Indicator Movement" is Carey (1981). This paragraph owes much to the WG 06 sessions on "Social Indicators", the XIIIth World Congress of Sociology in Bielefeld, July 18–23, 1994, and discussions with Dag Hareide (1990), a Norwegian NGO-activist and author of a book on *The Good Norway*.

housing, and some collective resources (public services, safety, etc.). The aim is not to measure what people experience or what happens to them; rather the focus is on the presumption that the material and social benefits provide different opportunities which in turn offer different qualities of lifestyle. This initially Scandinavian tradition attempts to develop a comprehensive set of statistics and is now used in all OECD countries.

• The measurement of social phenomena, indicating something about relationships between people. This many-sided tradition has roots in Continental European sociology. It is sometimes called misery research because of its focus on rates of crime, violence, suicide, drug and alcohol use, mental illness, societal erosion, etc.

The relevance and choice of social indicators depends on the social goals and performance being evaluated. Numerous sets of indicators have been assembled and are reported annually to complement economic data, for example the World Bank's annual World Development Report, the UNDP's Human Development Report, or UNICEF's annual State of the World's Children. Anderson (1991) highlights five predominant categories of social indicators: education and literacy; work and unemployment; consumption; distribution of income and wealth; and health.

COMPILATIONS OF LOOSE INDICATORS: CASE STUDY OF JACKSONVILLE, FLORIDA

The city of Jacksonville, Florida, USA, has taken the use of indicators one step further: they have developed a basket of indicators that not only measure elements of welfare and well-being, but do so in a way that is meaningful to local residents. Ordinary citizens are included in a process of determining local priorities of what to measure and in the actual measuring process—conducting interviews door-to-door. This community-oriented approach is interesting for its ability to incorporate a very democratic element into the establishment and application of welfare indicators in real politics.[3]

This project, started in 1985, is based on a strong motivation for

---

[3] See Hazel Henderson's extensive discussion of these issues in *Paradigms in Progress–Life Beyond Economics* (1991), especially Chapter 6, "The Indicator Crisis."

community improvement in the capital of Duval County, Florida. It is explicitly stated as the goal of both the Jacksonville Chamber of Commerce and the Jacksonville Community Council, Inc. to monitor the County's progress on an annual basis by means of selected representative quantitative indicators with a unique methodology of "interactive" participation by Jacksonville citizens.

The indicators include nine main categories comprised of 74 indicators on health, education, the natural environment, government/politics, the social environment, culture/recreation, mobility, public safety, and the economy with several subcategories. What is most relevant about the Jacksonville experience is that a democratic, local process was used to develop indicators that are meaningful to that community.

THE "HUMAN DEVELOPMENT INDEX" (HDI)

The Human Development Index (HDI) is presented by the UNDP as "an alternative to GNP for measuring the relative socio-economic progress of nations. It enables people and their governments to evaluate progress over time—and to determine priorities for policy intervention. It also permits instructive comparisons of experiences in different countries" (UNDP, 1994).

The HDI integrates three key components into a single composite index:

- Longevity —as measured by life expectancy at birth as the sole unadjusted indicator;
- Knowledge—as measured by two educational stock variables: adult literacy and mean years of schooling. The measure of educational achievement is adjusted by assessing a weight of two-thirds to literacy and one-third to mean years of schooling;
- Standard of living—as measured by purchasing power, based on real GDP per capita adjusted for the local cost of living (purchasing power parity, or PPP) with the underlying initial and innovative idea to produce a yardstick more comprehensive than GNP alone for measuring country progress, based on the premise of diminishing returns from income for human development: the higher the income relative to the poverty level, the more sharply the diminishing returns affect contribution of income to the HDI.

The breakthrough for one single HDI combination was to find a common measuring rod for the socio-economic "distance travelled" towards development. The HDI therefore sets a minimum and a maximum for each dimension and then shows where each country stands in relation to these scales.[4]

The HDI has been the subject of significant controversy because of questions over the methodology, consistency of data, and interpretation of its results (Kelley and Nohlen, 1990; Klingebiehl, 1992; Nohlen and Nuscheler, 1993). Nonetheless the HDI provides an important contrast to the traditional measurement of development only in terms of GNP or GNP per capita.

INDEX OF SUSTAINABLE ECONOMIC WELFARE (ISEW)

First published as a comprehensive appendix in the 1989 book *For the Common Good*, by Herman Daly and John B. Cobb, Jr. (1989), this ambitious index[5] is the most recent and comprehensive effort to construct an aggregate welfare index as a theoretical challenge to the GNP as a measure of welfare. Subsequently, this index has been derived for the UK (Jackson, 1994), Germany (Diefenbacher, 1994), the Netherlands (Rosenberg, 1995), Denmark (Jesperson, 1994), and Austria (Obermayr, 1994).

The ISEW approach begins with personal consumption as the basis of economic welfare. Personal consumption is then adjusted by an index of income distribution that gives greater weight to consumption when incomes are more equally distributed and less weight when incomes are less equal, relative to a given base year. Unlike the HDI, the ISEW does not reflect diminishing returns to consumption.

From this base, additions are made to reflect economic welfare derived from non-market labour production (i.e., household production),[6] government welfare inducing expenditures, and stocks of private goods and public infrastructure. Then the costs of current economic activity are subtracted: the costs of unemployment, commuting, automobile accidents, and water, air, soil, and noise pollution.

---

[4] For further technical details and a discussion of the pros and cons of such a procedure, see HDR 1994, p. 91 ff., especially 108 ff.

[5] In *The Green National Product*, the ISEW was reviewed extensively by a study group that elaborated the criticisms of the ISEW and the attempt to find a unified index in general (see Diefenbacher, 1994).

[6] While the ISEW does include household production, it does not so far include other nonmarket productive activities such as volunteer work.

Sustainability in an environmental context is then accounted for by subtracting the long-term costs of land-use changes from natural uses to urban uses, non-renewable natural resource depletion, fossil fuel and nuclear power use, and ozone layer depletion. Sustainability of the economic system itself is included by accounting for current consumption based on domestic capital growth and net foreign lending or borrowing.

The specific data that are included or excluded from the ISEW vary from country to country, partly as a result of data availability, partly because of different local priorities (i.e., environmental concerns), and partly because the methodology is still in a stage of development and experimentation. This makes it difficult to directly compare absolute measures of welfare between countries. On the other hand, comparison of the general trends gives insights into the economies that the GNP/GDP does not give.

The results of the ISEW analyses show that in almost all country studies so far, economic welfare per capita rose in the early parts of the study periods, and then declined in the later parts (except in the Austrian model, which moved roughly in parallel with GNP throughout the whole study period). The timing of the rise and fall varies by country. By contrast, in all countries, GNP per capita rose throughout the period, averaging about 2% growth per year from 1950 to 1990 in most industrialized countries. These trends are illustrated in Figures 9.2 and 9.3, which show the ISEW and GNP per capita for the UK and the GPI and GNP per capita for the United States, respectively. GPI means "genuine progress indicator." Cobb introduced the term GPI at the end of 1994 to replace the term ISEW because he found the former to be more accurate. The change is in name only, with the purpose of both avoiding debate over whether the index is a measure of welfare per se (which it is not intended to be) and focusing on the fundamental issue: building a better indicator of societal progress than the GNP.

Despite methodological differences, the ISEW's in many of the countries studied so far have demonstrated that the combination of social factors, income inequality, and environmental degradation have led to flat or declining welfare despite growing GNP,[7] as we have already mentioned in Chapter 1.

---

[7] In Germany, the ISEW rose from the late 1980s after having declined previously. This may be in large part due to the integration of the former East Germany and the heavy current investment there.

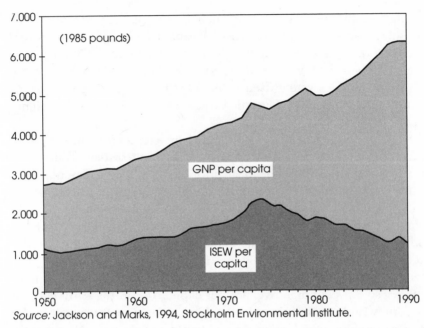

Source: Jackson and Marks, 1994, Stockholm Environmental Institute.

Figure 9.2 ISEW and GNP per capita, United Kingdom, 1950–1990

While the ISEW poses a significant theoretical challenge to the dominance of GNP, it is itself plagued by methodological questions. In particular, the centrality of consumption as the key to welfare is very important. Critics of the ISEW (and the GNP) argue that the ISEW, as a single index, is in itself too close to GNP to provide the information that society needs. Advocates of the ISEW generally agree with that judgement and argue that the value of the ISEW lies in its use as a rationale for policies other than the continuous support of growth. The ISEW can be viewed as an interim step towards both better accounting procedures and systems dynamics modeling efforts that will give better insights into the relationships between the social, environmental, and economic components of welfare and their relationship to sustainability. Daly and Cobb, writing about the original version of the ISEW, recognised that "the ISEW is far from perfect. All statistical measures are inevitably misleading in many respects. But the difference in policies ordered to the improvement of the ISEW and those ordered to the increase of GDP would be considerable, and they would help to buy time for the deeper changes that are needed" (Daly, 1989).

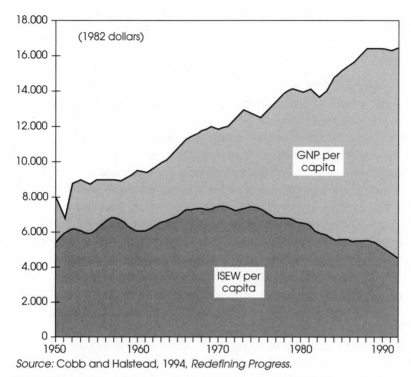

Source: Cobb and Halstead, 1994, *Redefining Progress.*

Figure 9.3   GPI and GNP per capita, United States, 1950–1992

The ISEW is also criticized by some for its ambitious scope. Ian Miles, a researcher at Manchester University, wrote, "Insurmountable difficulties are posed by the search for a single composite measure of welfare or quality of life. Perhaps we should be glad that human life remains too rich to be represented in one-dimensional terms" (Ekins, 1992).

Yet the argument for pursuing the ISEW concept remains strong, given the importance that the single composite index GNP holds in modern economies as a measure of welfare. Researchers argue that the difficulties described above are not insurmountable. More research is needed to refine the assumptions used in the ISEW regarding nonmarket values, especially of environmental impacts, and the incorporation of "sustainability" benchmarks. Further, Daly and Cobb, in addressing the criticism of the ISEW, respond that "whereas the conventional GNP accounting procedures ignore activities that are important but not easily measurable, the ISEW has the advantage of being generally accurate rather than precisely wrong" (Cobb and Cobb, 1994).

Some Remarks on Indicators for Sustainable Development[8]

The approaches described in the previous sections remain more or less in the realm of the well-being debate and welfare tradition, although the concept of sustainable environmental development is already to some extent integrated into the ISEW framework. In particular, the ISEW and other indicators in that genre do not track the specific cause-and-effect relationships related to natural resources. To have a better picture of the state of the environment itself and the causes of changes in environmental quality, more focused measures of the pressures, state, and responses to impacts on natural resources are needed to guide policymakers in developing environmental policy. In part due to the controversy surrounding welfare measurement, environmental measures were decoupled from the welfare debate in the 1980s, and the two fields—environmental indicators and welfare indicators—evolved largely separately.

The demand for environmental sustainability indicators has become especially prominent in research and policy since the UNCED-Conference in Rio de Janeiro in 1992. The key role of Sustainable Development Indicators (SDI) and the greening of the national accounts in measuring progress toward sustainable development has been defined as an integral element of the action plan known as "Agenda 21," the major achievement of the Earth Summit: "Commonly used indicators do not provide adequate indications of sustainability. Indicators of sustainable development need to be developed to provide solid bases for decision making at all levels and to contribute to a self-regulating sustainability of integrated environment and development systems."

As mentioned in Chapter 7 already, sustainable development includes economic, social, and environmental aspects. As a consequence, indicators for sustainable development imply the development of a framework that includes all three aspects.

Some additional remarks have to be made concerning the approach mentioned above, especially regarding the meaning of the sus-

[8] Both the author of this chapter and the editor are strongly aware of the growing relevance of sustainability or sustainable development indicators (SDI's). The focus of this report is on the necessity and possibilities of monetary valuation of environmental depletion and degradation. SDI issues have been dealth with in this section here, in the introduction to Part IV, in Chapter 14, and in Part V.

tainability indicators in realizing the Agenda 21 demand to implement "indicators of sustainable development." On an international level, it is especially the Department for Policy Coordination and Sustainable Development (DPCSD) of the UN division for Sustainable Development which is dealing with this issue. The DPCSD[9] is developing a "provisional core set of indicators for sustainable development for monitoring progress towards sustainable development at a national level through the implementation of Agenda 21" for the CSD (Commission for Sustainable Development) process. As an explicit demand of the developing countries additionally to the pure environmentally-oriented SDIs, this core set should include further issues: economic, social, and institutional indicators. In this respect, the SDIs offer the unique historical chance to recombine proper environmental issues and indexes with the socioeconomic ones. For further reading on this issue, we refer the reader to the forthcoming publications from the DPCSD.

ENVIRONMENTAL SUSTAINABILITY INDICATORS

Within the international scientific community, emphasis in the last few years has been put on the development of a framework to measure and reflect changes in the state of the environment. One of the prominent approaches in this field is the pressure–state–response approach (OECD, 1995). In this framework, pressure, state, and response are the primary categories in which data and indicators are organized to reflect the relationships between man and environment. The pressure category reflects attempts to measure the impacts from production and economic activity on the environment. Measurements are made in physical units, like kilograms of carbon emissions, or in amount of forest cover in a region, as measured by Graphical Information Systems. The state category is what we are most after: a measure of environmental quality. This is also the most difficult to monetarize or to capture in a single index or indicator. The response

---

[9] Together with other relevant organizations and some nongovernmental institutions and research institutes like the World Resource Institute in Washington, the Wuppertal Institute in Germany, the New Economics Foundation in the United Kingdom, and especially SCOPE (Scientific Committee of Problems of the Environment).

category resembles traditional economic accounting (SNA) most closely.

One has to realize that this approach, and other, complementary approaches like Natural Resource Accounting (NRA) and Environmental Resource Accounting (ERA), focus on the environmental dimension only. In this book, emphasis has been put on the flaws of the System of National Accounts and the necessity to improve it as an economic information system. Relevant approaches in this area are elaborated in Part IV and V.

But, as was stressed in the System of integrating Environmental and Economic Accounting (SEEA; see also Chapter 14), physical data and indicators are the building blocks essential for all monetary assessments. As a consequence, much attention is being paid to sustainability indicators from the perspective of economic and environmental accounting on a national level.[10] A future aim could be to integrate all these elements in the "Directions for the EU on the Environmental Indicators and Green National Accounting, the Integration of Environmental and Economic Information Systems." The following aim is explicitly stated:

> On the basis of a Report prepared by the Services, the Commission has developed a set of complementary actions establishing a European framework for "green accounting," which will provide:
>
> i. a European System of Integrated Economic and Environmental Indices (ESI), a much needed direct integration of economic performance and environmental pressures of economic sectors in a comparable way, within 2-3 years;
>
> ii. the larger and more fundamental work on the "greening" of the National Accounts in a satellite format (detailing environmental expenditures, establishing natural resource accounts, improving knowledge and methodologies for environmental damage assessment and monetary evaluation).

---

[10] The same holds on a company level for the environmental performance indicators (EPIs) which are being prepared in the new ISO-standard 14000 series (Environmental Management System). These EPIs create a whole new opportunity to make data and reporting statistically available. This opportunity is being researched by the Wuppertal Institute under the code name "micro-macro link" in the framework of the EU research project "Methodological problems in the calculation of adjusted national income figures," of which CBS the Netherlands is the project coordinator (Bosch, 1992–1996).

For further details, see "Directions for the EU on Environmental Indicators and Green National Accounting: The Integration of Environmental and Economic Information Systems (Com 94, 670 final, Brussels, 21.12.1994).[11]

CONCLUDING REMARKS

ECONOMIC AND ENVIRONMENTAL INDICATORS

In the international debate, experts tend to defend either the physical environmental indicators approach or the approach which aims at reflecting value changes of natural resources and environmental quality within the SNA. This debate is exaggerated. Many actors in this field do not seem to (or are unwilling to) understand that these approaches are not mutually exclusive. To the contrary, they are complementary and could be mutually reinforcing, as stressed by Stahmer in Chapter 14. They serve different purposes. *The physical environmental accounting approach aims at measuring and reflecting the state of the environment more accurately, whereas the SNA reform approach aims at improvement of the economic information system.* If a framework of social indicators would be added to these two dimensions, we would be well on the way of developing a more comprehensive framework for Sustainable Development Indicators.

---

[11] This document should be considered as a picture of the state of the art. It reflects the current EU views on the integration of economic and environmental accounting. The Club of Rome and WWF-International have their own vision on the sense of direction in this area. These visions are reflected in this book, the WWF-International report "Real Value for Nature" (Sheng, 1995), and the Action Plan "Taking Nature into Account," which has been presented at the international conference carrying the same name. This conference was held from May 31–June 1, 1995, in the European Parliament in Brussels. It was jointly organized by the European Commission, the European Parliament, WWF-International, and the Club of Rome. On this occasion, WWF-International and the Club of Rome also presented a document in which the common grounds and differences between the views of the European Commission and the European Parliament on the one hand, and the WWF/Club of Rome view on the other, have been made explicit. The conference and the related publications can be seen as a new phase in the process toward integrating economic and environmental accounting, in that the discussion on this issue has been carried beyond the relatively small group of experts in this field. It is on the political agenda now.

WELFARE INDICATORS

Various approaches have been proposed to improve the measurement of welfare. Both international agencies and driven individuals have put forward different types of approaches, which have contributed significantly to the ongoing process of the development of alternative indicators for societal progress.

A mainstream in this development cannot yet be identified. The various proposals have different characteristics in terms of objectives, levels of aggregation units, levels of democratic participation, etc. Until now, only a few initiatives have been taken to overview and assess the activities in this field. Though more conferences on ecological economics do pay attention to the indicator issue, to a large extent they serve only as a platform for experts to present (and defend) their specific methodologies. An exception to this rule is the seminar "Accounting for Change," which was organized by the New Economics Foundation in October 1994. This seminar brought together the key players in this field. Fundamental questions were raised not only to clarify the objectives and assess the methodologies, but also to assess the extent to which new indicators, when introduced, would influence decision making. The seminar report will probably provide a comprehensive overview and assessment of the most relevant issues in the indicator field.

C H A P T E R   1 0

The Role of Social Indicators

When we challenge the incompleteness of the GNP measure, we should be aware that for a number of years many efforts have been under way to produce new indicators. We can now take a closer look at this maze of new scorecards, statistics, and quality-of-life indexes which strive to redefine wealth and progress in order to change economic behavior.

Statistics determine our world view. We measure what we treasure. So statistics, however objective and accurate, are never value free but focus on what various societies deem to be important goals and values. Such statistical instrument panels, for better or worse, guide government, business, and individual decisions.

This new statistical alphabet, reflecting new concerns and goals, includes the United Nations Statistical Office's revised "Handbook of National Accounts" (1993) which offers EDP (Environmentally Adjusted Net Domestic Product); ENI (Environmentally Adjusted National Income); SNI (Sustainable National Income); and FISD (Framework for Indicators of Sustainable Development), as well as HDI (Human Development Index), published since 1990 by the United Nations Development Program. Other earlier US models include PQLI (Physical Quality of Life Index), MEW (Measure of Economic Welfare), both dating back to the 1970s; as well as ISEW (Index of Sustainable Economic Welfare), and the CFI (Country Futures Indicators).

These new scorecards and the greening of GNP/GDP national accounts partly reflect the new accounting in corporate balance sheets, reports, and books on environmental and social auditing.

Which of these new scorecards, accounting methods, and evaluation tools will emerge as benchmarks for investors, business, governments, and voters in the 21st century? At the very least, assumptions underlying old and new indicators are being clarified. The debate is both over what and how to measure, and what to do about values and amenities that are priceless. The World Bank's "World Development Report, 1993" focuses on health statistics and the need to invest in educating women. Vice president Ismail Serageldin advocates a triangular approach based equally on environmental, social, and economic indicators. Such thinking will guide the Bank's new Global Environment Facility's lending. (See Chapter 7.)

We are beginning to agree that society has been confusing means (i.e., GNP growth) with ends—human development and survival and further evolution under drastically changed planetary conditions.

Events are driving all of these changes. The speedup of global currency speculation forces the IMF to shift from per capita income denominated GDP to PPP (Purchasing Power Parity) in order to compare actual living standards. Waves of privatizations require adding capital assets into GNP to account for public infrastructure and investments—currently expensed as spending. A major issue concerns whether new indicators will be weighted in monetary terms to expand GDP, as for example the EDP and ISEW, or to unbundle the separate components—health, education, environment, etc.—so the public can follow their own concerns. A US survey in 1993 found that 72 percent of Americans preferred this approach.

On Earth Day, in April, 1993, President Clinton ordered the US Commerce Department's Bureau of Economic Analysis to correct the

US GDP to include environmental assets and costs. Much ferment about how to do this is summed up by the bureau's director, Dr. Carol S. Carson:

> Gross Domestic Product (GDP)—widely used around the world— is a measure of market oriented economic production. In turn, the level of production largely determines how much a community can consume. While the level of consumption of goods and services, both individually and collectively, is one of the most important factors influencing a community's welfare, there are many others—the existence of peace or war, technology, the environment, and income distribution, to name a few. Because these other factors do not enter into the measurement of GDP, additional measures are also needed to evaluate welfare or make policies regarding welfare. It would take a philosopher king to 'add' all the measures relevant to welfare into a single indicator useful for all times and communities; until then, a variety of measures will be needed along with GDP.

Today's debates concern how best to calculate the hidden costs of production passed on to taxpayers or future generations. Some are easy to quantify; costs of cleaning up pollution can be calculated and their increase marches in lock-step with the expansion of pollution control and environment industry sectors.

The difficulty with the upcoming indexes and the ways they will be measured is the discrepancy between the experts' opinions and public opinion. As Bartelmus says:

> EDP could be used to define sustainable economic growth in operational terms as: increases in EDP (which allows for the consumption of produced natural assets and the depletion and degradation of natural resources), assuming that the allowances made can be invested into capital maintenance and taking into account that past trends of depletion and degradation can be offset or mitigated by technological progress, discovery of natural resources, and changes in consumption patterns.

Such definitions still beg a host of questions about such single indexes: how will the public be told about such assumptions underlying EDP or how economists weighted all of these factors? How will balances be struck between tax-supported social programs dealing with family stress and illness that increase with rising unemployment due to corporate layoffs to bolster productivity and profits?

Sustainable development was defined in the Brundtland Report "Our Common Future," in 1987, as "development which meets the needs of the present without compromising the ability of future generations to meet their own needs." Thus, HDI shifts focus toward health, education, and fuller economic and democratic participation in people-centered development. The U.N.'s December, 1993 Expert Group Meeting on Sustainable Development Indicators targeted a new Framework of Indicators for Sustainable Development (FISD). The interest of social scientists and environmentalists in the need for overhauling the Gross National Product (GNP)/Gross Domestic Product (GDP)-based Systems of National Accounts (SNAs) began at least twenty-five years ago (Henderson, 1981 and 1991, Chapter 6). In the 1960s Emile van Lennep, former Secretary General of the Organization for Economic Cooperation and Development (OECD), had attempted to introduce social indicators into that organization's predominately economic analyses. Van Lennep encountered objections that such social indicators were "normative."[1] The best early examples of alternative indicators are the Index of Social Progress (ISP) devised by Estes, begun in 1974 and summarized in his "Trends in World Social Development: the Social Progress of Nations, 1970–1987" (Estes, 1988); and those developed by Morris for the Overseas Development Council of Washington, D.C., called Physical Quality of Life Index (PQLI). Neither received much attention at the time (Henderson, 1981, Chapter 13). The UNDP's public dissemination of all its annual Human Development Index editions since 1990 has created unprecedented levels of press attention. Wide interest in the Human Development Index (HDI) included much controversy. For example, its Human Freedom Index was criticized by many developing countries as being biased towards Western values. "HDI 1991" was also criticized because some developing countries saw its human right policy focus as being used as another set of new "social and environmental conditionalities" that the World Bank and Northern bankers could add to their loans (already replete with familiar conditionalities such as structural adjustment of their economies).

Other global indicators have received less attention. The International Monetary Fund's "World Economic Outlook, 1993" received

---

[1] Even though, of course, economic indicators are also normative.

even more attention than the HDI when it issued an Addendum which converted conventional per capita GDP rankings to Purchasing Power Parities (PPPs) to correct for currency fluctuations. This re-ranking via PPPs catapulted China to the rank of the world's third largest economy after the US and Japan.

It is not enough only to point to the effects and symptoms of the existing unsustainable global economic and geopolitical system (e.g. desertification, pollution, poverty, injustice). It is necessary also to highlight the causes and the structural barriers to achieving ecologically sustainable, equitable human development, which include: an inequitable global economic order, reinforced by the Bretton Woods financial institutions, such as the International Monetary Fund (IMF), and the General Agreement on Trade and Tariffs (GATT); as well as unequal terms of trade, dominance over the U.N. by the G7 countries, and the global arms race which still accounts for some one trillion dollars annually. This accounts for the usefulness of the HDI's indicators of military-to-civilian budget ratios (see Figure 10.1) and the work of Sivard and her Washington-based World Military Expenditures Indicators (Sivard, 1991). Another new indicator is that of the US-

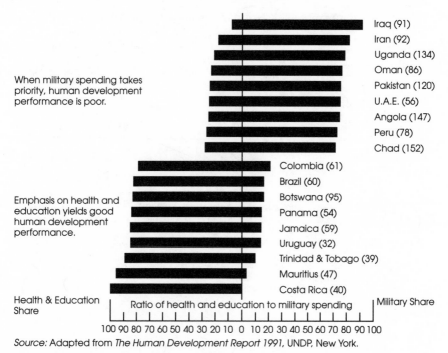

Figure 10.1 Military spending and human development performance

based World Game, which relates global arms expenditures to global needs (see Figure 10.2). Examples of how HDI's indicators reveal long-term patterns that are glossed over by per capita averaged UNSNAs include HDI's highlighting of widening poverty gaps in its 1992 report and its focus in the 1993 edition on the ominous syndrome of jobless economic growth now prevalent in all the world's economies.

Where are the social and structural leverage points pushing the new indicators agenda? A bottleneck concerns neoclassical economics and its divergence from other disciplines. It clearly is a body of thought that came into being to rationalize and support the goals of the industrial revolution as it began in Britain 300 years ago. Adam Smith merely described what he saw emerging with unabashed approval, in that a way of producing social progress, as he saw it, could at last be founded on human frailties—acquisitiveness, greed, self-centeredness, and competition—rather than waiting until human beings perfected themselves (Henderson, 1981, Chapter 8).

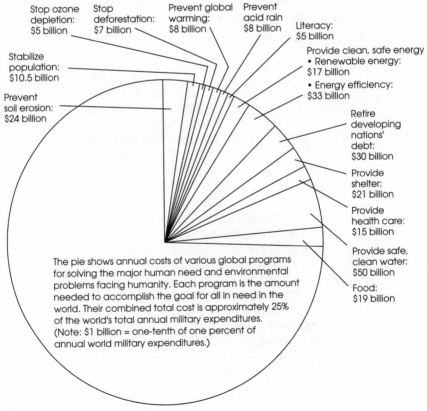

Figure 10.2   What the world wants

These are the categories through which Jacksonville, Florida, has chosen to measure its quality of life. Statistical data has been recorded in these categories since 1983.

The economy
(includes individual well-being and community economic health)

Total unemployment rate
Black unemployment rate
Teenage unemployment rate
Effective buying income per capita
Annual retail sales per capita
Average price of previously owned, single-family home
Revenue generated by the bed tax
Taxed taxable value of real estate
Cost of 1,000 kw hours of electricity (DEA)
New housing starts

Public safety
(includes the perception of public safety, and the quantity and quality of law enforcement, fire protection, and rescue services)

Index crimes per 100,000 population
Percentage who have been victims of a crime
Percentage who feel safe walking alone at night
Rescue call response time
Fire call response time
Police call response time
Motor vehicle accidents per 1,000 population
Fire and rescue operating expenditures per capita
Law enforcement operating expenditures per capita

Health
(refers to the physical and mental health of residents and the local system of health care)

Age-adjusted deaths per 100,000 population
Infant deaths per 1,000 live births
Deaths due to heart disease per 100,000 population
Suicides per 100,000 population
Deaths due to cirrhosis of the liver per 100,000 population
Packs of cigarettes sold per capita
Percentage who exercise three times per week
Percentage who rate the medical health care system good or excellent
Percentage who rate their own health good or excellent

Education
(includes the system of public education (kindergarten through 12th grade), higher education, including adult education, and the overall literacy and educational attainment of the population)

K-12:
• Standard Achievement Test scores
• Student dropout rate
• Educational expenditure per student
• Average public school teacher salaries
• Percentage of teachers holding advanced degrees
Higher education:
• Percentage of faculty holding terminal degrees
• Average faculty salaries at public institutions
• Total student enrollment
• Academic degrees awarded
Educational attainment:
• Percentage who are high school graduates or above
• Percentage who are college graduates or above

Natural environment
(includes the natural elements of the earth's ecosystem, the quality and quantity of water, air, green space, and landscaping and visual aesthetics)

Days when the air quality index is in the good range
Frequency of compliance of St. John's River with water quality standards
Frequency of compliance of tributaries with water quality standards (desolved oxygen)
New sewer permits issued
Sign permits issued
Environmental public employees per 100,000 population
Per capita tons of solid waste

Mobility
(refers to opportunities for people to travel freely within Jacksonville and between Jacksonville and other locations)

Total weekday commercial flights in and out of JIA
Direct flight destinations to and from JIA
Average weekday ridership on JTA buses
Average miles of JTA bus service per weekday
Commuting time from downtown along J. Turner Butler/I-95
Commuting time from downtown along Atlantic Blvd
Commuting time from downtown along San Jose Blvd
Commuting time from downtown along Roosevelt Blvd
Commuting time from downtown along I-95 North

(continued)

Figure 10.3   Life in Jacksonville: Quality indicators for progress
Source: Jacksonville Chamber of Commerce, Jacksonville, FL.

Government/politics
(includes an informed and active citizenry, profession-
alism and performance of local government)

Percentage of population eighteen and over regis-
tered to vote
Percentage of registered voters who voted
Percentage of households purchasing local Sunday
newspaper
Percentage who can name accurately two City Coun-
cil members
City capital outlay expenditures per capita
Total City revenues per capita
Percentage of City Council members who are black
Percentage of City Council members who are female
Rate of Planning Department/City Council concur-
rence on rezoning petitions
Percentage who rate the quality of local government
leadership good or excellent
Days from arrest to disposition of criminal cases

Social environment
(encompasses collective or group concerns such as
equality of opportunity, racial harmony, family life, human services, philanthropy and volunteerism)

Child abuse and neglect reports per 1,000 children
under 18

Employment discrimination complaints filed with the
Jacksonville Equal Opportunity Commission
Percentage of white persons who believe racism to be
a problem
Percentage of nonwhite persons whe believe racism to
be a problem
City government human services expenditures per
capita
Contributions per capita to the United Way and its
member agencies
Percentage who volunteered time during the past year
Licensed daycare spaces per 1,000 children under 5
years

Culture/recreation
(includes the available supply and use of sports and
entertainment events, the performing and visual arts,
public recreation, and leisure activities)

Annual expenditures of major arts organizations
Public parks acreage per 1,000 population
City parks and recreation facilities per 100,000
population
Public library materials per capita
Public library book circulation per capita
Days of bookings of major city facilities
Zoo attendance per 1,000 population

Figure 10.3   *(Continued)*

Progress in developing indicators of sustainable development will
demand that such broader indicators be interdisciplinary. Gauging the
progress of complex societies and their many dimensions of quality
of life using a single disciplinary approach, is, on the face of it, absurd.
Multidisciplinary approaches include such indicators as Jacksonville,
Florida's Quality Indicators of Progress, operating since 1983 (see Figure 10.3). The city of Seattle and many others are modeling new sustainable city indicators on those of Jacksonville, while the World
Health Organization reports on numerous "Healthy City Indicators"
worldwide. National statistical needs of developing countries are also
multidisciplinary, since all countries have their own cultural DNA
codes (diverse goals and values) around which their own indicators
can be designed (Caracas report, 1990). The South Commission took
this approach in its report, "Challenge to the South" (1990). Thus,
sustainable development indicators will diverge methodologically in
many ways from traditional national accounting. Such economic techniques, being normative, non-transparent, and unaccountable to elec-
torates, seem to preempt the very processes of democracy. The fears

"Beyond money-denominated, per capita averaged growth of GNP"

Re-formulated GNP to correct errors and provide more information:	Complementary indicators of progress towards society's goals:
• Purchasing Power Parity (PPP): corrects for currency fluctuations • Income distribution: is the poverty gap widening or narrowing? • Community based accounting: to complement current enterprise basis • Informal, household sector production: measures all hours worked (paid and unpaid) • Deduct social and environmental costs: a "net" accounting avoids double counting • Account for depletion of non-renewable resources: analogous to a capital consumption deflator • Energy input/GDP ratio: measures energy efficiency, recycling • Military/civilian budget ratio: measures effectiveness of governments • Capital asset account for built infrastructure and public resources (many economists agreed this is needed; some include environment as a resource)	Population: birth rates, crowding, age distribution Education: literacy levels, school dropout and repetition rates Health: infant mortality, low birth weight, weight/height/age Nutrition: e.g. calories per day, protein/ carbohydrates ratio, etc. Basic services: e.g. access to clean water, etc. Shelter: housing availability/quality, homelessness, etc. Public safety: crime Child development: World Health Organization, UNESCO, etc. Political participation and democratic process: e.g. Amnesty International data: money influence in elections, electoral participation rates Status of minority and ethnic populations and women: e.g. human rights data Air and water quality and environmental pollutions levels: air pollution in urban areas Environmental resource depletion: hectares of land, forests lost annually Biodiversity and species loss: e.g. Canada's environmental indicators Culture, recreational resources: e.g. Jacksonville, Florida

Figure 10.4   Country Future Indicators™

Copyright © 1989 Hazel Henderson

of many South policymakers, mentioned earlier, are thus understandable: that HDI and green indicators are too normative, like all such measurements, in the sense that they focus on what societies hold is important.

The main weakness of HDI is precisely that, so far, it still uses statistical methods of trying to aggregate diverse elements, using often traditional weighting to come up with some analog of GNP/GDP. Daly (1990) also acknowledges this problem with his Index of Sustainable Economic Welfare (ISEW), illustrating his concern with a variation on the slogan Carleton cigarettes used in the USA: "If you smoke, please try Carleton," i.e. since all countries are already using indexes such as GNP/GDP, at least they should try something marginally better. However, in principle Daly, Sachs, and other economists concerned with sustainable development feel that societies should not use such overall indexes.

To address this problem the Country Futures Indicators (CFI) (see Figure 10.4) are unbundled so as to be transparent and multidisciplinary and accessible to the public. They also try to overcome another correction on the social front: e.g. unpaid work, poverty gaps, etc. Many efforts have been made or are under way to design such indicators, such as Repetto's in valuing forest resources in Indonesia and Costa Rica; Costanza's work on the value of wetlands (based on Odum's net energy analyses at the University of Florida); and case studies by Lutz, El Serafy, Pezzey, von Amsberg, and others at the World Bank. By definition, they must be viewed as a partial aspect of sustainable development indicators, as these try to encompass the whole spectrum of human well-being.

Consequently, we try to forge methodological links with holistic, social approaches like those of the HDI, the Society for the Advancement of Social Economics, and Canada's Green Indicators, to name a few.

The overview in this chapter is not all-comprehensive. Rather than evaluating and assessing all the efforts done or under way, our intention is to give a description, in order to make it widely known that the search for new indicators or indexes is now a global one.

PART FOUR

ENVIRONMENTAL

ADJUSTMENT OF THE SYSTEM

OF NATIONAL ACCOUNTS

Introduction

After elaborating the historical roots of economic growth (Part I), the impact of current levels and patterns of economic activity on the environment and the quality of life (Part II), and the process of sustainable development (Part III), in Parts IV and V we will deal with GDP and the System of National Accounts (SNA). In Chapters 2 and 3, the historical background of the development of the SNA and the societal impact of one of the main aggregates, GDP, were discussed. The failures of GDP to reflect economic progress have also been elaborated. In addition, the required improvement of indicators for sustainable development has been discussed in Chapter 9.

In this part, the focus is narrower. While indicators for sustainable development are concerned with economic, environmental, and social aspects, we now focus on one of the *economic* indicators: GDP. The reason for this has been discussed in Chapter 9 already: the dominant role of GDP as an indicator for economic progress. The remedy for this disease is twofold: 1) improvement of GDP as an economic indicator, and 2) returning GDP to its proper context: GDP is only one of the economic indicators, which should in turn be seen in the context of environmental and social indicators.

From here, we shall concentrate on the integration of environmental aspects in the System of National Accounts in general, and calculation of GDP in particular. In this context, systems for natural resource and environmental accounting and methodologies to value changes in natural resource stocks and environmental degradation are relevant.

Much work has been done in the field of natural resource and environmental accounting. The World Resource Institute has carried out important case studies for Costa Rica and Indonesia (Repetto, 1989). From 1983 to 1988, UNEP and the World Bank organized a series of workshops in which most relevant actors participated and which resulted in the publication of *Environmental Accounting for Sustainable Development* (Ahmad et al, 1989). In 1993, UNSTAT and the World Bank organized a symposium on environmental accounting and published *Towards Improved Accounting for the Environment* (Lutz, 1993). This publication contains the latest developments in the field of conceptual and methodological issues, a summary of the U.N. System for Integrated Economic and Environmental Accounting, and comments on it by several experts. It also contains the results of two recent case studies, for Papua New Guinea and Mexico. Recently, the *U.N. Handbook on Integrated Environmental and Economic Accounting* was published (United Nations, 1993), providing guidelines for a general environmental accounting framework for national statistical offices. The 1993 SNA itself contains a separate section on integrated environmental-economic satellite accounting.

In recent years, many national statistical offices have gained experience with environmental accounting, mainly regarding the systematic recording and linking of physical data on stocks and flows of natural resources and emissions to air, water, and soil to monetary data on economic activity on the sectoral and national level as recorded in the System of National Accounts. Environmental accounting in physical units provides important information on the use of natural resources, the pressures on (and the state of) the environment. However, physical environmental accounting as such does not enable policy makers to make a tradeoff between economic and environmental gains and losses and obviously does not solve the fundamental flaws in the calculation of GDP. The *U.N. Handbook on Integrated Environmental and Economic Accounting* acknowledges this. As a consequence, it states that physical data are needed to describe the state of the environment and to provide a solid basis for any attempt to value the changes of natural assets in monetary units. The central question is: How to value the observed and recorded changes?

A lot of work has been done on developing methodologies to value natural resources and the environment in monetary terms. Most academic textbooks on environmental economics contain a description and assessment of familiar valuation methodologies such as the Contingent Valuation Method, hedonic pricing, and the travel-cost method.

The Contingent Valuation Method (CVM) is the most prominent approach. By means of questionnaires or gaming situations people are asked to indicate the amount of money they are willing to pay to maintain or improve (a specific element of) the natural environment. People may also be asked how much they should receive as compensation for a certain loss of environmental quality. Thus, a hypothetical, contingent market is created and people are enabled to express their preferences. Aside from having such advantages as technical applicability and the direct recording of values expressed by participants, the Contingent Valuation Method has several disadvantages. One example is the difference between the so-called willingness to pay and the amount of money that people are actually prepared to pay in practice. Another is that the willingness to pay depends on the income of people in the first place. Besides these drawbacks, most people have only a very limited knowledge of environmental issues and the state of the environment. This makes it difficult to assess the true value of the values expressed.

In the *hedonic price approach,* the changes in the value of property like land and houses resulting from differences in environmental quality are recorded. The change in the value of the property serves as an estimate for the value of changes in environmental quality.

In the *travel-cost method,* the benefits of a specific site are estimated, for instance a recreational center. The (extra) time people are willing to spend travelling to the site and back, the travel expenses themselves (fuel) and the relation between the travelling time and the time actually spent at the site are the main variables. A further extension could be to estimate the amount of money people are willing to pay to maintain or improve the site.

For an excellent elaboration of CVM, hedonic pricing, and the travel-cost method, see Pearce and Turner (1990).

These valuation methods provide insights which can be used to support more integrated economic and environmental decisionmaking, but mainly in rather specific situations. Case studies have been carried out to estimate the recreational value of lakes, forests, or the property value of houses, buildings and land. It is clear that the described methods are object-, project-, or micro-oriented. Another important feature is the focus on people's revealed preferences. This can be an advantage in the sense that it is a more or less democratic approach. The direct involvement of people can also be a disadvantage, however, for reasons such as those mentioned earlier (limited knowledge and the introduction of bias).

From the above it may be easily understood that these valuation methodologies cannot readily be applied in the context of environmental adjustment of GDP and the System of National Accounts, which is sectoral and macro-oriented. The estimation of value changes in the quality and quantity of natural assets requires other, meso- and macro-oriented valuation methodologies.

Part IV presents the work of experts on the main topics relating to environmental adjustment of GDP. Chapter 11 deals with the discussion on so-called environmental defensive expenditures in relation to GDP. For several years already, there has been an ongoing debate on deduction of these expenditures from GDP. Environmental defensive expenditures are meant to restore or avoid environmental damage but some categories are accounted for as value added. In this Chapter, Leipert's view on this matter is presented. Chapter 12 deals with El Serafy's view on the depletion of natural resources in relation to national income and GDP. Serafy's argument is based on Hick's definition of income and explicitly uses the weak sustainability criterion as its point of departure. He also gives some arguments against the application of the net price method, another market-based valuation method which has been used by the World Resource Institute in the Indonesian and Costa Rican case studies. Serafy has developed the user cost method, which aims at estimating the value-added element (true income) and the user-cost element (depletion costs), starting from GDP. The net-price and the user-cost approach are two examples of market-based valuation methodologies. Chapter 14 concentrates on the efforts of Hueting and his colleagues to estimate the national income in a hypothetical, environmentally sustainable economy. One of the basic assumptions in their approach is that in society there is a preference for sustainable development. As a consequence, one can set sustainability standards instead of constructing a demand curve (which is not possible, according to the authors) and calculate the costs of strategies to achieve these standards. Finally, Chapter 15 details the vision of Stahmer, who contributed significantly to the *U.N. Handbook on Integrated Environmental and Economic Accounting.* Besides summarizing the proposed extensions of the SNA framework concerning integrated economic and environmental accounting, Stahmer has been asked to write an assessment of the similarities, differences, strengths and weaknesses of the available proposals for valuation methodologies in the context of GDP and the System of National Accounts.

CHAPTER 11

Defensive Expenditures

The traditional indicator of economic growth transforms a failure into a success. GNP rises in the wake of a production process that disregards the environment, and it rises still further when the given environmental damages are mitigated through economic activities. The expenditures associated with these counteractions at all economic levels are generally called compensatory or defensive expenditures.

In this chapter we concentrate on the question, What are (environmentally) defensive expenditures? We are interested in the theoretical implications of this concept and in the empirical findings on environmentally defensive expenditures (EDE). After discussing this matter, we move on to the problem of how defensive expenditures are to be defined and measured in the context of an environmentally adjusted national income figure.

WHAT ARE DEFENSIVE EXPENDITURES?

The National Accounts concepts of GNP and of economic growth are undiscriminating. All apparently final monetary economic activities are added to the GNP regardless of their purpose and the function they play in production and consumption. The aggregation of monetary economic activities into GNP is doubtless useful as an aid to stabilization and fiscal policy, which requires information about the level and development of market production, income distribution, and consumer and investment expenditures—independent of their contribution to quality of life. However, this one-dimensional concept of GNP is clearly inadequate when the aim is indeed to calculate a measure of true net income—net of all costs of keeping capital (incl. natural capital) intact. To an (ever) increasing extent, GNP contains transactions that as such cannot be given a positive value, and whose sole function is to repair damage and avoid environmental burdens brought about by the negative effects of economic activities (defensive expenditures).

Defensive expenditures comprise those economic activities with which we defend ourselves against the unwanted side-effects (negative external effects) of our aggregate production and consumption (see, e.g., Olson, 1977 and Daly, 1989). They are understood as expenditures to cure, neutralize, eliminate, avoid, and anticipate burdens on and damage to the environment (and living conditions in general) caused by the economic process in industrial societies.

Defensive expenditures are not superfluous in the short term. Under the given economic and technological conditions they are both necessary and useful. Expenditures on environmental preservation and restoration fulfill positive functions here and now. First, their causal relationship allows the cost nature of these expenditures to become clear: the deterioration of certain environmental and living conditions is historically attendant on industrial production, the market outputs of which are registered unaltered in the GNP. However, the expenditures that compensate for these problems or attempt, in view of the serious risks, to prevent them, are additional monetary expenses made to achieve the conventional bundle of goods-and-services-plus-environmental services that were formerly cost-free or nearly cost-free.

EDE arise with the transition from the environment as a free good to the environment as a scarce good. With the growing environmental burdens and natural resource needs of the economic growth process in the industrial world, it became increasingly costly to obtain the normal productive and consumptive services of the environment we were accustomed to having (nearly) cost-free in the past. At the beginning of that new "era" of scarce environmental services, the seventies, we had to pay mainly for purifying polluted air and water and for reducing the negative effects of noise emissions. In the eighties we were forced to pay additionally for cleaning up polluted soils; for purifying contaminated ground water; for meeting stricter emission standards for air pollutants, waste water, and waste disposal; for repairing damages to buildings and works of art; and for treating damaged forests.

Since the end of the eighties we have reached another, higher step on the ladder of costs and are now also paying for substituting ozone-depleting CFCs, for mitigating the greenhouse effect, for extending technological and financial transfers from the North to the South, for securing biodiversity in terrestrial and maritime ecosystems, and for developing new, more environmentally friendly systems for transport and chemical and agricultural production.

Seen from a dynamic perspective, defensive expenditures are additional economic costs incurred by a specific pattern of growth and development. Parts of the final production indicated in the National Product are not outputs, but inputs, i.e. production and consumption costs. The additional defensive expenditures can be interpreted as a real income transfer from the human production system to the environment. They are comparable with the real income transfer from the oil-consuming countries to the OPEC states after the oil price explosions of 1973/74 and 1979/80. These real income transfers amount to a loss of income that cannot be spent once again for consumption or investment.

This situation is still not perceived as such by the public in the industrial countries, however. In the bargainings of distribution, the environment is not seen as an actor with a legitimate claim to a certain fraction of the (growth of) production in exchange for the valuable services with which it provides the economy. The price increases induced by EDE are not distinguished from other causes like pure inflationary forces, and they normally lead to higher wage and income demands, which then compensate for such price increases. Identification, measurement, and regular publication of an environmentally adjusted national income could help to sensitize the public to the cost nature of EDE in our new world of scarce environmental services.

The concept of defensive expenditures demonstrates that in certain cases less can be more, and vice versa. The development of certain environmentally detrimental, material- and energy-wasting, and geographically centralized economic structures necessitates additional compensatory expenditures to achieve unaltered economic, social, and environmental goals. A change in these types of cost-raising economic structures could lead to a reduction in (compensatory) expenditures without a reduction in the quality of life.

A CONCEPT OF ENVIRONMENTALLY DEFENSIVE EXPENDITURES

The elaboration of a suitable concept of EDE should start with a wide concept of environmentally related expenditures which attempts to be all-inclusive:

1. Treatment and compensation of damages caused by environmental burdens (e.g. repairs to buildings, production facilities, historical monuments, compensatory treatment of health impairments, treatment of damaged forests, additional cleaning activities).

2. Restoration and clean-up activities (cleaning up of toxic waste dumps and polluted production sites, treatment of polluted surface and ground water by water works, restoration of spoiled and destroyed ecosystems and landscapes).

3. Evasion and screening activities to avoid damage from environmental burdens (e.g., noise emissions).

4. Environmental protection, abatement and disposal activities in specific environmental protection facilities (elimination and treatment (including recycling) of pollutants, waste water and toxic waste before they are discharged into the environment).

5. Process-integrated technologies; energy- and resource-saving technologies; environmentally sound products.

6. Activities for changing prevailing environmentally damaging consumption and production patterns in an environment-friendly direction.

The most comprehensive concept of EDE would entail categories 1-5. Categories 4 and 5, and partially 3, are aimed mainly at avoiding the emission of additional pollutants to the environment, whereas 1 and 2, and partially 3, comprise mainly economic activities that compensate, repair, and restore vis-à-vis environmental losses and environmental, economic, cultural, and health damages originating in the past. The activities under 6 and, partially, 5 can be interpreted as being proactive, anticipatory, and structure-changing expenditures that help to render superfluous repair, treatment, and abatement activities in the mid- and long-term future. An (absolute and relative) increase in these kinds of proactive expenditures is to be interpreted as the success of an economic policy that integrates environmental dimensions.

Most environmentally related expenditures are still *defensive* expenditures. The desired change in the structure of these expenditures is primarily from the extremely defensive fractions 1 and 2 to the milder defensive fraction 4 and, in another step, from 4 to 5 and 6.

The national and international systems of environmental statistics currently in use or under development are oriented mainly towards categories 4 and 5 and sometimes to parts of 2. In other words, they generally cover only a fraction of total EDE. An exception is the Draft Classification of environment-related Defensive Activities in the *SNA Handbook on Integrated Environmental and Economic Accounting* (United Nations, 1993, p. 41), which aims to cover the whole spectrum.

EMPIRICAL EVIDENCE ON DEFENSIVE EXPENDITURES: THE EXAMPLE OF GERMANY

According to empirical calculations for the (former) Federal Republic of Germany there has been an absolute and relative increase in both main categories of EDE (Leipert, 1989):

- environmental protection and disposal expenditures; and
- repair, treatment, and cleaning-up expenditures; resulting from environmental burdens and damages.

The total economic burden to society brought about by EDE increased in West Germany, in relation to GNP, from 1.5 per cent in 1970 to 3.4 per cent in 1988.[1] While, in nearly twenty years, total GNP increased by 50 per cent, total EDE rocketed by more than 300 per cent. The absolute figures are given in Table 11.1.

At the end of the 1980s nearly one twenty-fifth of the production included in GNP is being consumed to prevent, dispose of, and compensate for, the environmental burdens and damages that have followed from the negative environmental impacts of the economic growth process.

The empirical evidence of the last few years suggests that this trend has continued to date and will become probably even more pronounced in the next ten years or so. Rapid growth of preservation, recycling, disposal, clean-up, repair, and restoration activities seems to be preprogrammed for the next ten to fifteen years. Since the second half of the 1980s the new, generally stricter environmental standards introduced earlier in the decade have led to rapidly increasing environmental protection investments by industries and municipalities. The dynamics of this growth are reflected in a characteristic report in the daily press, according to which the chemical company BASF is to invest twice as much in environmental protection as it has in the previous ten years—DM 2 billion, compared with 1 billion.

Current expenses for using the accumulated stock of environmental protection and treatment facilities are also increasing from year to year, with growth rates of 10 per cent and more. Desulphurization equipment for coal-fired power plants, the building of a "secure" waste disposal infrastructure (controlled landfills for toxic waste, recycling, separation, and incineration facilities) and so on are expensive, but they also entail major long-term current expenses that will greatly inflate the total operating costs of environmental protection in industry and in the public sector.

---

[1] An exact figure could not be calculated because many quantities comprise both final deliveries and intermediate goods and services which are difficult or impossible to break down. It is therefore better to speak of a ratio (not an exact percentage) between defensive expenditures and GNP. If it is taken into consideration, however, that the present calculations are based chiefly on minimal values and that a number of other categories of defensive expenditures could not be computed, it can be assumed that the genuine proportions of the GNP have at least the same dimensions.

Table 11.1

ENVIRONMENTAL DEFENSIVE EXPENDITURES IN THE (FORMER) FEDERAL REPUBLIC OF GERMANY, 1970–1988, IN BILLIONS OF DEUTSCHMARKS AT CONSTANT PRICES (CALCULATED WITH THE DEFLATOR OF FINAL DOMESTIC USE, WITH 1980 AS THE BASE YEAR)

	1970	1980	1985	1988
A. Environmental protection expenditures				
1. Environmental protection expenditures by industry (investments, current expenditures, and purchase of services)	5.70	8.90	14.0	19.3
2. Environmental protection expenditures by the public sector (investments, current expenditures)	6.00	12.80	11.80	15.00
3. Additional costs for the installation of catalytic-converters in cars	—	—	—	1.00
4. Additional costs for improving the security standards of nuclear power stations	—	—	2.30	2.30
Total of A	11.70	21.70	28.10	37.60
B. Expenditures for compensating unavoided environmental damages				
1. Additional costs due to noise pollution	—	2.00	2.00	2.00
2. Costs for cleaning up hazardous waste sites	—	0.05	0.15	1.00
3. Costs for asbestos removal	—	—	—	0.40
4. Costs for PCB-condensor replacement	—	—	—	0.10
5. Pollution-related treatment costs of waterworks	1.60	3.30	4.80	5.00
(a) Pollution of surface water	0.11	0.14	0.15	—
(b) Pollution of ground water	—	—	0.94	1.00
(c) Removal of chlorinated hydrocarbons	—	—	0.50	1.00
(d) Removal of nitrates				
6. Costs of water transport induced by polluted ground water	1.00	1.00	1.00	1.00
7. Costs of cleaning up ship discharges and oil spills	—	—	0.10	0.10
8. Additional costs in forestry due to air pollution	—	—	0.40	0.50
9. Costs of anti-erosion measures	—	—	0.30	0.50
10. Additional costs incurred by damage to buildings, materials, equipment, and textiles	2.40	2.40	4.00	4.00
11. Additional costs for restoring works of art	—	0.05	0.15	0.15
12. Additional costs for car owners incurred by salt on roads	—	2.55	2.85	3.10
Total of B	5.10	11.50	17.20	19.85
Total of A and B	16.80	33.20	45.30	57.45
For comparison: GNP at market prices	1,134.00	1,485.20	1,578.10	1,699.40

Source: Leipert, 1989, p. 218ff.

ENVIRONMENTAL DEFENSIVE EXPENDITURES AND ENVIRONMENTALLY ADJUSTED NATIONAL INCOME

ENVIRONMENTAL DEFENSIVE EXPENDITURES AND NET ENVIRONMENTAL DAMAGE

Annual calculation of EDE as an absolute figure and as a ratio to GDP provides important economic and political information for the industrial countries. It should be published regularly by their statistical bodies, as well as by the supranational organizations EU and OECD. However, this figure is not defined in such a way that it can be used unchanged in calculating an environmentally adjusted national income. This paragraph elaborates the role of EDE in the calculation of an environmentally adjusted national product figure.

Generally speaking, the EDE of today cannot maintain the scale, quality, or diversity of the natural assets. Without the activities financed by EDE, the degradation and depletion of natural assets would have been greater than is actually the case.

We need to account for the net environmental damages and for the net deductions of natural assets that we observe. Consequently, the total economic cost of attaining the standards of sustainability in the current period ("keep capital intact") are the actual maintenance "investments" (actual defensive costs) plus the monetary costs of net degradation and depletion of natural assets.

EDE AND AVOIDANCE COSTS

EDE are expenditures which have actually been made during the accounting period. Potential avoidance costs, on the other hand, are hypothetical expenditures that should have been made in order to keep natural capital intact during the accounting period or to achieve a certain environmental standard. If properly defined, the concept of EDE and the so-called avoidance costs are related. Defensive expenditures first avoid actual emissions of pollutants and then repair or treat the damages and burdens generated by emissions in earlier periods. A parallel situation occurs with potential avoidance costs: a distinction can be made between avoidance measures directed towards pollutant flows in the current year (avoidance of actual emissions) and those directed towards treating stocks of pollutants (restoration of damages generated by emissions in the past). In this sense the avoidance costs relating to emissions in the accounting period reflect the reduction in environmental quality that has not been prevented by actual environmental defensive expenditures.

EDE IN RELATION TO GDP

In this section, a few conceptual considerations will be dealt with briefly.

First, EDE are functionally comparable with replacement investments—compensating for the wear and tear of man-made capital—which should be deducted from GNP to arrive at the Net National Product.

Second, national income is a flow measure. Therefore, only a flow-oriented concept of EDE fits in with a concept of environmentally adjusted national income. Because GNP is a flow measure, only that part of EDE should be deducted that avoids or compensates for the decrease of environmental quality generated by the production and consumption flow of the accounting period. The research question here is: how to solve the periodization problem between actual and past-oriented defensive expenditures?

For our purposes here the various items of EDE should be differentiated according to the following criteria: What are (mainly) actual defensive expenditures directed towards (potential) emission flows in the accounting period; and, what are (mainly) past-oriented defensive expenditures in reaction to (the negative effects of) stocks of pollutants originating from emissions in previous periods?

Third, for the purposes of adjustment, the cost figures rather than expenditure figures of defensive activities are required. In the ensuing calculations, depreciation rather than investment values must be employed. In some cases, especially in the case of large strategic investments by business, it is not easy to make a distinction between the costs relating to conventional strategic investment and the purely environmentally oriented costs. Clear criteria to make this distinction still have to be developed.

WHICH EDE SHOULD BE DEDUCTED?

The depreciation rates of the industrial and municipal environmental protection capital stock are already incorporated in the general depreciation figures for produced man-made capital. This type of defensive capital costs is dealt with in the transition from GDP to NDP.

Not included in this transition, however, are such cost items as the environmental services of consultants and engineering firms, defensive expenditures of households and current expenditures on environmental protection in the municipal and industrial sectors.

The expenses of running the environmental protection capital stock in the industrial sector, which are reflected in National Accounts as intermediate consumption, are not accounted for as value added in the first place. What must be avoided is (negative) double counting by deducting them a second time. According to Leipert, those expenses in the industrial sector which are reflected in the economic output figure should be deducted.

The running costs of environmental preservation in the municipal sector are included in GDP. They should be identified and deducted from NDP. However, insofar as the services are financed by fees paid by the industrial and household sectors (as users of dwellings, the latter are treated as part of the business sector), they are not entered as final expenditures in GNP. Here again we must avoid double counting.

There remain two categories of EDE which are currently erroneously accounted for as value added and should be deducted from GDP: 1) the value-added component of environment-oriented services of firms for other firms and the public sector; and 2) environmentally defensive purchases by private households. Examples of the first category are the environmental services of consultants and engineering firms for private and municipal firms. Examples of the second category are the extra expenditures on health care for problems due to air pollution, for example, and the extra expenditures on cars equipped with catalytic converters.

CONCLUDING REMARKS

Politicians, advisers, and ecologically conscious scientists working for an environment-oriented change in economic policy need a measure of economic production/growth that is deflated by the environmental costs induced by that production/growth.

This task is politically (and also scientifically) so important that we have to accept imperfect solutions. We must teach the public that the currently prevailing production, income, and growth concepts and calculations are only seemingly objective, seemingly valid, and seemingly "modern." They benefit from the fact that the produced goods and services are valued at market prices which seem to be self-evident. Crucially, though, the market prices of today are determined with essentially no allowance for the external environmental costs of production and consumption and are therefore increasingly far removed from the economic-ecological "truth."

The assumptions behind the calculation of an economic production/growth figure are "conventions." Their conventional character will be more evident in the future, since a calculation of an environmentally adjusted national product figure lacks the self-evidence of the current accounting systems based on market prices. Together with this change, however, comes a corresponding increase in the complexity of the economic discourse in the era of environmental scarcity. The collective/public value of economic (private/microeconomic/business) activities depends on the environmental requirements and environmental effects of those activities. The societal uncertainties implied by this new phenomenon cannot be overcome simply by monetarizing the environmental losses by means of an apparently correct valuation method.

An environmentally adjusted national income (net national product) entails making due adjustments for the environmental costs of production and income spending, which are not yet being accounted for. The criterion for calculating annual net national income, "Keep Capital Intact," must be extended with an additional capital dimension: "scarce environmental potentials/assets." This extended cost accounting of the annual production/returns on income is an important step to a timely welfare interpretation of the current environmentally deteriorating style of production and consumption. However, construction of such an accounting system is not intended to serve as a comprehensive measure of economic welfare in the tradition of the "Net National Welfare" line of thought.

Leipert opts for a calculation of environmental costs based on actual defensive expenditures and potential avoidance/substitution costs. National product/income is a flow measure. We should therefore deduct only those environmental costs induced by production/consumption activities in the current period:

GDP ÷ depreciation of manmade capital (including environmental
         protection capital stock)
   = NDP ÷ environmental defensive costs (actual replacement
         investments in the natural capital stock)
   ÷ potential replacement investments in natural capital (potential
         avoidance costs/monetarization of the net loss of natural capital)
   = EcoNational Product
   = Environmentally adjusted national product

The calculation of the real environmental losses (avoidance costs) should focus on the greatest and most urgent problems. The first attempts to arrive at an environmentally adjusted national product figure should not aim to be all-inclusive and comprehensive from the outset. It will be a process of trial and error. Initial estimates will be more akin to a model calculation. With time we will gain the experience to develop the "right" conventions for calculating the environmentally adjusted national product/income figure.

CHAPTER 12

Depletion of Natural Resources

INTRODUCTION

All natural resources are depletable in the sense that they are capable of being destroyed unless care is exercised to maintain them, or at least not disturb them. But there is a subcategory of natural resources that has traditionally been considered as depletable, exhaustible, or non-renewable—which is the subject of this chapter.

Depletable resources are those that cannot be renewed or regenerated, including fossil fuels, other minerals, unreplenishable aquifers, and the like. Such resources are essentially exhaustible even if some recycling is possible so that a certain reversibility of their depletion is feasible.[1] Living resources, such as forests, though capable of being regenerated, can still be depleted, either inadvertently or deliberately when mined by their owners to extinction or near extinction.[2]

Similarly, the ability of nature to assimilate the wastes associated with production and consumption represents a resource with a virtually given absorptive capacity that is also depletable, although in some cases reversing depletion may be possible. The ability to reverse the depletion of a renewable resource should improve with technological progress unless, of course, the resource has already become totally extinct.

The focus in this chapter, however, will be on that class of marketable resources conventionally viewed as depletable or exhaustible, with minerals representing a prime example of such resources.

CONVENTIONAL ACCOUNTING

Depletable natural resources that are exploited in the public sector have been treated as a gift of nature, carrying no resource cost to their exploiters in the sense that no provisions were made through fiscal or other means to earmark part of the revenue from exploitation for productive investment in order to sustain prosperity after the resource has been exhausted. This was virtually the method endorsed by the 1968 version of the United Nations System of National Accounts (SNA) which has been in use in most countries until now. With the 1993 revision of the SNA, a new approach (or approaches) is proposed to be elaborated in so-called Satellite Accounts for the environment. Where depletable resources were being worked in the private sector, there has been a tendency to take depletion into account since a decline in reserves as a result of exploitation obviously diminished the value of the exploiter's assets. Tax authorities in some countries permitted a depletion allowance to be charged against gross profits for the reckoning of net profits. To the extent that the depletion allow-

---

[1] Though the laws of thermodynamics indicate that matter cannot be destroyed, once a material like gasoline is burnt, the products of the combustion cannot economically be reconstituted back into gasoline, given present technology.

[2] Population pressure and severe poverty can drastically raise time preference (that is, valuing the present much more than the future) and force people to liquidate natural capital in order to survive.

ance was correctly estimated, and exploitation was carried out in the private sector, the national accounts came out right. In the majority of developing countries, however, where natural resources have been worked in the public sector, proceeds from mining natural resources have been treated as income. The faster the depletion, the more prosperous the country would seem to be and the more rapid its apparent economic growth. The fact that such prosperity would be ephemeral, and that the apparent growth was misleading, did not seem to worry most economists who continued to base their country analysis and policy prescriptions uncritically on the erroneously reckoned national accounts.

LEVELS OF SUSTAINABILITY

Better national accounting can provide a first step towards attaining sustainability. A country that derives a significant part of its economic prosperity from depleting its natural resources would be living beyond its means if the proceeds from liquidating those resources were treated as current income. Correcting the accounts so that asset sales are not conflated with income would impart the message that the economy concerned is poorer than originally perceived; that it is on an unsustainable path; and that, inevitably, current consumption levels will have to decline unless remedial action is taken.[3]

One may distinguish here among three ascending levels of sustainability: weak, strong, and absurd sustainability.[4] The sustainability to which better accounting should lead is weak sustainability. The argument for seeking weak sustainability is entirely economic, and is embedded in the correct measurement of income which must take account of capital deterioration. Capital here is fundist capital as defined by Hicks. The late Professor Sir John Hicks (1974) drew attention to the difference in perception of capital between two schools of economists: the fundists, who thought of capital as a fund or value that represents all assets; and the materialists, who thought of capital as concrete categories of machines, buildings, inventories, etc. Main-

---

[3] Remedial action will often include devaluing the national currency and raising the saving and investment ratios.

[4] Serageldin, Daly, and Goodland distinguish *four* levels of sustainability: weak, sensible, strong, and absurdly strong sustainability. See Chapter 7.

taining fundist capital intact would mean that a decline in any one category of capital does not matter as long as another category is increased so that total capital remains the same. This concept of (fundist) capital is the one consistent with weak sustainability.

Strong sustainability builds on weak sustainability and aims, not just at maintaining total capital intact, but at maintaining subcategories of capital separately, particularly natural capital. This, of course, is a very restricting constraint. The argument is usually presented as developing from a lack of substitutability between natural capital and other forms of capital. For instance it is argued by advocates of strong sustainability that fish cannot substitute for fishing vessels, and timber cannot substitute for sawmills. But this argument would deny the possibility that final wants are capable of being changed so that fish may be replaced by other forms of proteins and timber by other materials. Ultimately, however, diminished substitutability will assert itself, particularly at the global level.

According to the concept of strong sustainability, natural capital deterioration must be compensated for by fresh investment in renewable natural capital substitutes. In other words, if petroleum is extracted, at least part of the proceeds (see below) should be set aside and reinvested in renewable natural energy sources, such as sugar cane, not just in any assets. The fact that scarce water and soil resources needed to produce sugar cane may not themselves be sustainable is seldom recognized.

Absurd sustainability would aim at leaving the environment, including marketable natural resources, untouched, and this represents an extreme conservationist objective.

The sustainability being considered in this chapter is weak sustainability which belongs entirely to positive (or analytical and descriptive) economics. For this purpose, national income and product accounting plays a leading part.

Various Objectives

Once it is recognized that the environment is indispensable as a source of raw materials and energy, as well as a waste-sink for the emissions of economic activities, maintaining the environment can be appreciated as essential for sustaining economic prosperity. A standard tool of assessing economic prosperity is the estimation of national income and product, which would give misleading readings if environmental deterioration were left out (El Serafy, 1991).

But not all those who seek the adjustment of the national accounts to reflect environmental change have the same objective.

An aim frequently claimed for adjusting the accounts is comprehensiveness. Conventional national income estimates do not include many activities such as illegal and underground transactions of an economic nature. Nor do they cover domestic work or personal services performed by household members. Also human capital change including improved skills by education, and losses due to aging and mortality is conventionally omitted from the national accounts. But there are valid reasons why such changes are left out, whereas declines in marketable natural resources should be covered.[5] National accounting cannot be all-comprehensive, and accounting for environmental change will always be partial. Much environmental change will remain difficult or even impossible to value meaningfully in money terms, and this should be accepted.

Another frequently stated aim of several writers on the subject is to record the loss of welfare resulting from the deterioration of environmental amenities, from pollution and other resource degradation. While not underrating this objective, the primary concern of the economist should be to correct the misleading accounts for the purpose of better economic analysis. Welfare considerations are dealt with in Chapter 10 and Part V of this report.

Yet another objective is to estimate wealth changes from year to year. This objective corresponds to many environmentalists' concern for showing periodic deteriorations in the state of the environment,

---

[5] Reasons for excluding such activities vary, and the new United Nations System of National Accounts, in proposing satellite accounts for the environment, aims at gradually extending satellite accounts to activities hitherto left out. One overriding consideration of the accountant, which applies especially to depleting resources, is to record losses when they occur, even without actual transactions taking place, while refraining from recording gains until they are actually realized. This derives from concern that if unrealized gains are certified as income, this may lead to higher consumption, so that if prices were to drop later, the intactness of capital would be threatened. As to changes in human capital, there are valid reasons why these should be left out of income accounting. If investment in skills is productive, it will inevitably reflect itself directly in improved productivity, and therefore higher income, without the accountant having to make any special adjustment to the accounts. All this leads to the conclusion that even partially corrected accounts would still be better than accounts which are not corrected at all, and therefore convey misleading information about the economy and its progress.

expressed in money terms. Not content with physical indicators of such changes, the notion of putting a value on the stocks is thought to bring home to society the extent of environmental deterioration. Concern with valuing the stocks threatens to undermine the value of income accounting for the environment. Because of the importance of this point, it is considered in detail below. Income accounting is of a much greater use to the economist than wealth accounting, and trying to cover both in an integrated system would sacrifice income reckoning at the altar of dubious and (to the economist) unnecessary estimates of wealth. Even the stock of human-made capital is rarely, if ever, valued comprehensively in a national accounting framework.

The need to adjust the accounts for environmental change will vary from country to country depending on the importance of the environmental change to sustaining economic activity in the country concerned. Priority, however, should be given to accounting for changes in marketable natural resources which tend to play a much greater role in the economies of the developing countries than in those of the industrialized nations.

The overriding consideration should therefore be to aim at adjusting the conventional estimates for what is economically important and can be also meaningfully covered, in the expectation that as time goes by the adjustment to the national accounts will expand to cover hitherto uncovered aspects and thus will grow in comprehensiveness. But it should be realized that no system of accounts can ever achieve complete comprehensiveness.[6] Thus, focusing the adjustment on a few resources that are significant for sustainability in specific country settings is totally defensible. This is the approach which was pioneered by the World Resources Institute in practical studies for countries like Indonesia and Costa Rica (Repetto et al., 1989; Solorzano et al., 1991).[7]

---

[6] This may be compared to Pigou's handling of welfare economics. Pigou considers economic welfare as part of human welfare, which the economist is not particularly qualified to analyze. To Pigou, economic welfare was that part of human welfare that could be brought into a "relationship with the measuring rod of money." Starting with that part of the environment that could be brought into a relationship with the measuring rod of money, the adjustment to the accounts might expand over time to include more and more aspects of the environment (El Serafy, 1993).

[7] Focusing on a few natural resources that are important for the economy concerned is praiseworthy, but the method used in the Indonesia study for adjusting the accounts for depletable resources is debatable; in particular it erred by integrating reassessments of petroleum reserves directly into income measurements.

DEFINITION OF INCOME

Income, or more precisely, net income, is by definition sustainable. In order to estimate net income generated in a specific period for an individual, a corporation or a nation, capital consumption or depreciation has to be recognized and accounted for. In an accounting sense, enough of current receipts should be set aside and devoted to maintaining capital so that the ability to generate future income is not impaired. If enough is set aside for this purpose, the income owners may spend their net income on consumption in the knowledge that, other things being equal, their future consumption will not deteriorate. If the owners aspire to a higher level of future consumption, net capital formation must be positive. This means that not all net income should be consumed, but enough of it be devoted to investment in order to achieve the desired rise in future consumption.

Traditionally, national income accounting has paid attention to the maintenance of human-made capital, recognizing capital consumption, or depreciation, as a legitimate deduction from gross income before an estimate can be made of net income which determines the maximum that can be consumed in the accounting period if future income generation is not to be undermined. It is this net income which is true or sustainable income, not the gross income which is unsustainable. Despite this, gross income as estimated in the national accounts—whether in national (GNP) or domestic (GDP) terms—is regularly used for international comparisons, and for gauging economic growth. This is because the adjustment for national or domestic capital consumption tends to be rough and imprecise, and for many countries it is not even made at all. Besides, it is thought that if capital consumption is ignored, its presumed limited size will not fault international comparisons based on the gross product, and probably will not affect much the inferred economic growth rates, provided that temporal consistency is maintained in the measurements.

INCOME AND VALUE ADDED

Correcting gross income for depletable natural resources is quite different from adjusting it for depreciation. Whereas depreciation of human-made capital may be presumed to lie within some typical proportional limits, natural resource depletion has no norms: it can be negligible, or it can be considerable. It can affect one resource or several resources at a time. So an investigation is needed to assess its magnitude.

The fundamental concept underlying income and product estimation is value added. The periodic income received by members of an economy derives from the creation of value that is additional to the value of the inputs used in the process of production. Value added is generated by the services of factors of production (principally labor and capital) as they are applied to nonfactor inputs. From this value added the factor owners, i.e. the workers and proprietors of capital, derive their income. In fact it is the sum-total of value added in all the productive units within an economy that makes up the domestic product realized during the accounting period. When natural resources, even those that are commercially traded, are being depleted and the receipts from this depletion are regarded as value added when in fact they are mostly or partly the proceeds from asset liquidation, the accounts violate the fundamental principle that income must come from value added. In other words, what passes as income would be contaminated with capital elements that should be totally excluded from it. Uncorrected GDP that contains such capital elements would be wrongly measured, and labeling it gross does not eradicate the error.

Depletable resources, as argued earlier, would normally present few problems in national income estimation if they were all appropriated, and their exploitation were carried out in the private sector. Where such resources are known, and the market sets a price on them, the areas in which they are located would command a market price reflecting their subsoil contents. And as exploitation of the deposits proceeds, with progressively less subsoil deposits remaining, the value of the site diminishes. For a private enterprise, what comes out of the soil by way of periodic exploitation cannot all be regarded as profits (or value added) until a depletion allowance is reckoned and excluded from the proceeds. But, for the majority of developing countries where depletable resources are being exploited in the public sector, no depletion allowance is normally estimated, and the proceeds from asset liquidation get treated as current income in the national accounts, and in the fiscal accounts as current revenue; and, if internationally traded, as current account receipts in the balance of payments—even though they can clearly be unrepeatable exports of a capital nature.

Accounting for Depletable Resources: A Balance Sheet Concept

It seems that a popular way to account for depletable resources is to treat such depletion as depreciation or capital consumption in parallel with human-made capital. This method conforms essentially with the approach proposed under the new SNA, though the latter has attempted to avoid some, but not all, of the contradictions inherent in such an approach.

There are several reservations about treating depletion as capital consumption to be taken out of an unadjusted gross product in order to reckon the net product. First, this will leave the gross domestic product unadjusted, focusing instead on adjusting the net product. For most countries, however, the net product is not at all estimated. Second, and more fundamentally, the unadjusted gross product will contain proceeds from asset liquidation which do not represent value added, the fundamental concept underlying GDP. Third, the analogy with human-made capital breaks down because human-made capital consumption is fairly limited as a proportion of GDP while depletion of natural resources can be substantial and varies greatly with the rate of exploitation in relation to the size of the reserves. Fourth, if attention is shifted, as it should be, from the gross to the net product as the indicator of sustainable income, the results do not make much sense from the point of view of income measurement, as illustrated by the following example.

Imagine the extreme case of a Saudi Arabia, where all GDP comes out of a depletable resource whose cost of extraction is nil. If the proceeds from exploitation are 100, GDP, as conventionally estimated, would be 100. And since this very amount represents the decline in the stock of the depletable resource, taking it out of GDP as depreciation would indicate a net domestic product of exactly zero. In other words the sustainable income of such a country would be nil (Case 1 in Table 12.1).

The above calculation, it will be noted, rests on the assumption that the price of the resource remains constant between the beginning and the end of the year. The problem is further complicated if the price of the resource changes, which it normally does. A popular procedure is to value opening stocks (i.e. the value of the reserves as of January 1 at the price prevailing in the market at that date; and value the closing stocks at the prices prevailing on December 31). Assume that the price increases by 20 per cent (from 1.00 to 1.20) gradually

Table 12.1

"NONSENSICAL" ADJUSTMENT

	Stocks (tons)	Extraction (tons)	Price ($)	Conventional GDP ($)	Value of stock ($)	Depreciation (change in stock) ($)	Conventional NNP (GDP + change in stock) ($)
Case 1							
Jan-01	10,000		1.0		10,000		
		100	av.:1.0	100		−100	100 − 100 = 0
Dec-31	9,900		1.0		9,900		
Case 2							
Jan-01	10,000		1.0		10,000		
		100	av.:1.1	110		+1,880	110 + 1,880 = 1,990
Dec-31	9,900		1.2		11,880		
Case 3							
Jan-01	10,000		1.0		10,000		
		100	av.:0.9	90		−2,080	90 − 2,080 = −1,990
Dec-31	9,900		0.8		7,920		

in the course of the year so that, roughly speaking, the unadjusted, conventionally estimated GDP rises from 100 to 110. If opening stocks were 100 times annual extraction, the adjustment of the gross product (now 110) would be a positive 1,880,[8] so that the net product would be 1990 (Case 2 in Table 12.1). If on the other hand there occurred a drop of 20 per cent in the price, the gross product would be approximately 90, and the net product a negative 1,990 (Case 3 in Table 12.1). This is clearly an untenable result.[9]

REASSESSMENT OF RESERVE STOCKS

Quite often the stocks of a depletable resource are reassessed, and found to be larger or smaller than was previously believed. The reassessment of stocks has presented the national income accountant with an apparent dilemma. Clearly if a balance sheet of environmental assets is to be drawn, such reassessments would appear in the wealth accounts of society. But should changes in the balance sheet be automatically reflected in the flow accounts so that income estimates must be adjusted by the full change in the balance sheet in the year the change occurs?

The pioneering study of the accounts of Indonesia made by the World Resources Institute integrated the stock reassessments into the estimation of the net product, thus producing nonsensical adjustments. The 1993 revision of the SNA would, while recording stock reassessments in balance sheets, keep them separate from the flow accounts. The new SNA, however, claims to offer an "integrated" approach, but integration obviously does not extend to reconciliation between the stocks and flow accounts.[10]

---

[8] Depreciation can be positive, but it is normally negative.

[9] The accounting convention regarding the valuation of end-period stock is to use the opening-period price or the current price whichever is less—a rule of which some economists writing on the subject do not seem to be aware.

[10] According to El Serafy, the "integrated" approach of the new SNA may be taken to mean an integration (in satellite accounts) of environmental changes into the traditional accounts which are being referred to by the United Nations Statistical Division (UNSTAT) as economic accounts, as if the environment had no bearing on economic measurements! The new *SNA Handbook on Integrated Environmental and Economic Accounting* (United Nations, 1993) thus seems to emphasize the wrong notion that the environment is extra-economic. See Bartelmus, P., E. Lutz and J. van Tongeren (1993).

While the new SNA appears to be half-hearted about reflecting stock reassessments in the flow accounts, impounding such reassessments in Reconciliation Accounts whose status is not clearly indicated, the United States has gone a step further and integrated the reassessment of stocks fully into the estimates of the domestic product. For the first quarter of 1994, the adjustment made for depletable resources in the national accounts of the United States followed the simplistic approach of fully deducting the decline of stocks on account of exploitation from the gross product, and adding to the latter the upward reassessment of reserves in the quarter, thus reaching the dubious conclusion that the accounts have been adjusted for depletable resource changes, and that the adjustment turned out to be negligible, since the addition to stocks through re-estimation offset the decline due to exploitation (US Dept. of Commerce, 1994).

OTHER APPROACHES

Other approaches to accounting for depletable resources include two notable examples. First, there is the approach that seeks accounting in physical units without (at any rate initially) attempting to integrate these in income or wealth measurements. The efforts made in France to track changes in the physical stocks of natural resources in patrimonial accounts represent such an approach which remains ambitious in its aims as it is (as yet) far removed from yielding operationally useful results. Secondly, in the Netherlands a two-track approach is being pursued, one to develop physical indicators of stock changes, and another to adjust national income for such changes using proposals made by Hueting who is more interested in assessing welfare than in positive economic measurements. Hueting has developed the concept of environmental functions that get lost or eroded, and whose cost of repair (to restore them to their original levels or to attain socially defined standards) should be charged against income generation. Hueting suggests shadow pricing these functions where the market is not helpful, but the shadow-pricing process he proposes is not always operational. Applying his approach to depletable resources, Hueting recommends estimating the costs necessary to substitute a depletable resource at a future date by a renewable alternative source.

However, one should realize that predicting future economic and technological development introduces an element of uncertainty in the estimation of the costs of environmental losses, which rightly is one of the criticisms of this approach (Hueting, 1993; see also Chapter 14).[11]

THE USER COST APPROACH

In a paper published by El Serafy in 1981, a method was proposed to adjust the national income of countries that rely on depletable resources in order to gauge their true or sustainable income (El Serafy, 1981, 1989). This method is based on the concept of a user cost of the depletable resource, on the argument that extraction has an inherent intertemporal opportunity cost deriving from the depletability of the resource; by extracting a barrel of oil today it will not be available tomorrow.

This user cost corresponds to what Alfred Marshall called a mining royalty, to be distinguished from Ricardian rent which, by definition, derives from the indestructible powers of the soil. The sum-total of royalty and rent makes up the surplus a mine owner derives from the exploitation of a depletable resource over and above the costs directly incurred for working the mine. Many analysts loosely call this surplus rent, which thus gets confused with Ricardian rent. Ricardian rent, on account of the indestructibility of the powers of the soil, is true income, whereas the royalty or the user cost is a capital element that should not be confused with income (Marshall, 1920). Marshall, following Ricardo, has stressed the difference between mine rent and farm rent viz:

> . . . when a vein had once given up its treasure, it could produce no more. This difference [between farms and mines] is illustrated by the fact that the rent of a mine is calculated on a different principle

---

[11] Hueting is on the whole skeptical of discounting future flows.

from that of a farm. The farmer contracts to give back the land as rich as he found it: a mining company cannot do this; and while the farmer's rent is reckoned by the year, mining rent consists chiefly of royalties which are levied in proportion to the stores that are taken out of nature's storehouse.[12]

The user cost method for the estimation of depletion cost in mining activities treats the reserves as a stock (or inventories) which can be drawn during the accounting period in toto or in part out of the ground depending on the owner's entrepreneurial decisions. Whether the exploitation profile is economically optimal or not (see below) is irrelevant. Even when the owner's decisions regarding how much to mine in any one period turns out in retrospect to have been misguided, the accountant has, nevertheless, to account for whatever decisions were made. The user cost method however, is sensitive to the discount rate used. This discount rate is inevitably an arbitrary rate, but can be changed periodically to reflect changes in the prospective returns expected from investing the equivalent of the user cost so that it may yield future income that could be sustained even after the resource has been totally exhausted. Taking the cue as mentioned before from Hicks's chapter on Income in *Value and Capital,* where he suggested that income derived from a wasting asset may be estimated by converting the proceeds into a permanent income stream (Hicks, 1946), El Serafy explored Hicks's proposal further. Making a few assumptions while applying accounting procedures, and treating reserves as inventories and not as fixed capital, El Serafy estimated X (or true income) as a proportion of R (the receipts from a wasting asset net of extraction costs) with the aid of only two parameters: n (the life expectancy of the resource at the current extraction rate) and r (the expected real rate of return on investing the user cost, $R - X$, in material or financial assets at home or abroad in order to sustain the same income level X even after the resource has been totally exhausted).

---

[12] It is interesting that Marshall in his *Principles of Economics,* Ch. 2, quotes Ricardo himself as making this distinction *viz:* "The compensation given (by the lessee) for the mine or quarry is paid for the value of the coal or stone which can be removed from them, and has no connection with the original or indestructible powers of the land."

> **True income, X as a proportion of net proceeds R, can be estimated as follows:**[13]
>
> $$X/R = 1 - \{1/(1+r)\}^{n+1}.$$
>
> The rest of R (which, as a ratio, is the complement of X/R) represents the user cost which should be totally excluded from the gross product on the argument that it is not depreciation, but a quantity expressing the using-up of inventories, and no further adjustment of gross or net income would be required to cover depletion.[14]

Since several analysts have persisted in treating depletion of natural resources as depreciation rather than as using-up inventories, one could argue that the user cost is the right quantity that should be used for "depreciation," but certainly not the total decline in the value of the reserves which, as already explained, would wipe out the activity altogether from the net product. Nevertheless, El Serafy has continued to insist that treating depletion as depreciation involves a conceptual confusion.

There are several advantages to employing the user cost method in order to account for depletion of exhaustible resources. First, it will always indicate a downward adjustment of conventionally calculated income even if new reserves are located. In fact, reserve discoveries and re-estimation can be integrated directly in the adjustment through the changed life expectancy of the resource. Second, the user cost method will not lead to the nonsensical result of zero net income, as explained above. Third, the method is flexible, being rooted in the accounting process which deals with income measurements one year at a time; if conditions change, the parameters used for the calculations can be altered.

However, it is extremely important to estimate the discount rate appropriately. For instance, when the user cost approach was applied

---

[13] The derivation of this formula and the underlying assumptions are stated in El Serafy (1981).
[14] See Table 12.2 for some illustrative calculations of user cost values. This method met with the approval of Hicks (personal communication to the author in 1987). It has been the subject of much analysis and application. See, for instance, Hartwick et al. (1993).

in the case studies for Mexico and Papua New Guinea (respectively by van Tongeren et al., and Bartelmus et al.: see Chapters 6 and 7 in Lutz, op. cit.) where an inappropriate discount rate (r) was used. The r in the El Serafy formula is a precautionary rate representing a target yield to be earned on investing the user cost to create a future stream of perpetual income. While a real rate of five percent has been suggested by El Serafy as (an optimistic) guide, a ten percent rate was used in the studies cited above, which resulted in drastically underestimating the user cost (See Table 12.2). This has led to the erroneous judgment that the user cost method has been tried in these studies and indicated little adjustment in income on account of depletion.

As the last column of Table 12.2 shows, the user cost (which is the adjustment that should be made to GDP) is significantly reduced when a discount rate as high as 10 per cent is employed. This is, as stated earlier, the rate used for adjusting the national accounts of Mexico and Papua New Guinea when the user cost method was applied. At a 10 per cent discount rate and a life expectancy of 25 years, the user cost is only 8 per cent of sale proceeds, whereas it is no less than 60 per cent when a discount rate of 2 per cent is used. At the risk of repetition, the discount rate that should be used for the calculation of the user cost is not an ambitious target of yield to be sought from

Table 12.2

USER COST AS PERCENTAGE OF SALE PROCEEDS

Life expectancy years	Discount rate		
	2%	5%	10%
2	94	86	68
5	89	75	56
10	80	58	35
15	73	46	22
25	60	28	8
50	36	8	1
100	14	1	0

Source: El Serafy (1989).

future investments, but a realistic and precautionary yield in real terms that can reasonably be expected from the investment. As stated before, when in doubt, the accountant almost instinctively leans to the side of caution: underestimating income rather than overestimating it. A high discount rate would contradict the logic of the process which tells the owner of the resource that a large enough part of sale proceeds should be saved for the benefit of sustaining current income into the future.

APPLICABILITY TO RENEWABLE RESOURCES

The same user cost approach, as argued earlier, is perfectly applicable to renewable resources. If exploitation of such resources is kept within their regenerative capacity, then no income adjustment will be called for—what comes out of the resource is income. But where exploitation exceeds this capacity and results in mining the renewable resource, a user cost for any such mining should be estimated by the same method, to be reinvested so that the income from the activity can be sustained.

POLICY IMPLICATIONS

Accounting for income is of great importance, and wrong accounting can do enormous harm to the understanding of the economy concerned, and hence can mislead policy recommendations. This is particularly relevant to developing countries where natural resources tend to contribute significantly to their economies. Counting asset liquidation as growth is obviously wrong. It exaggerates income and thus encourages unwarranted consumption, thereby discouraging saving and investment. It can lead countries to undertake too much external debt with later disastrous implications. It hides fundamental disequilibria in the balance of payments, and also conceals many of the symptoms of the Dutch Disease phenomenon which, in resource-depleting economies, upsets the domestic terms of trade between tradables and non-tradables and thus obstructs sustainable development.

ACCOUNTING AND ECONOMICS: CONCLUDING REMARKS

National income measurements straddle the disciplines of economics and accounting. In the current critical phase of reviewing the methods used for national income accounting, experts are bringing to the discussion the perspectives of their differing specializations, and there is much confusion in what is being written about the subject. In time, such confusion is bound to recede in the light of debate and informed discussion. It is worth stressing, however, that accountants and economists normally have different perspectives, with economists tending to be forward-looking, and inclined to view resources as a medium for optimization of decisions. They concern themselves with such aspects as pricing, and market forms, as well as with ideal rates of resource depletion. Accountants, on the other hand, hardly ever look forward. They are in effect economic historians, seeking to describe, in accounting language, what had taken place in a period that has already ended. Keeping capital intact has remained a pillar of accounting methods since the profession was born. Capital and its depreciation are not easily measured, and have presented the accountants with problems of estimation no less formidable than those that have confronted the economists. This is because the life of capital, by definition, extends well beyond one accounting period, and the accountant is forced to look forward to ascertain how long a piece of capital will last. More so than the economist who, with the help of assumptions, is able to build airtight compartments within which elegant analyses can be conducted, the accountant has to produce rough-and-ready answers that must serve the practical purpose of guarding against capital consumption. Capital, after all, is needed for the sustainability of future income, and where there is doubt, the accountant would rather err on the side of caution, underestimating income rather than exaggerating it. In doing this, the accountant plays an invaluable role as a guardian of sustainability.

Lastly, a word is perhaps needed in defense of the weak sustainability that underlies much of the argument in this chapter. Certainly many environmentalists would favor a stronger sustainability, and the present writer may agree with them, but finds the medium of national income accounting unsuitable for this purpose. For in order to sustain income into the future, capital, in whatever form, should be kept intact. The notion that specific forms of capital should be formed to replace the depleted assets is extraneous to the process of national accounting which is an ex post process and has nothing to do with prescriptions about future entrepreneurial or social decisions.

Those writers who have advocated the reinvestment of the user cost in projects that would generate sustainable alternatives to the resource being exhausted, arguing that human-made capital and natural capital are not perfect substitutes, are saying something that is beyond national accounting. Besides, many economists would argue that substitutability between the two forms of capital does indeed exist although it becomes more and more difficult as substitution proceeds and natural capital gets scarcer and scarcer. The limitation imposed by consumer income forces all of us to effect substitution in the final markets, and this should get transmitted back to the markets for inputs. While this latter argument is theoretically valid, its weakness in practice stems from the fact that the markets often fail to indicate natural resource scarcities properly. In this case direct public policy action is needed to maintain and enhance natural capital, but this task cannot be handled adequately through the medium of national accounting. Not all aspects of the environment, however, are capable of being captured in the national accounts which have to be carried out inevitably in money terms. National income accounting can account for income, but cannot properly account for the entirety of the environment. For the latter task, physical indicators are needed to complement income accounting.

CHAPTER 13

Estimating Sustainable National Income

A YARDSTICK FOR ECONOMIC DEVELOPMENT

The importance of a good yardstick for economic development is self-evident. With reference to one or just a few indicators, society can then tell whether there has been economic progress, decline, or stagnation in a given year. Ideally, such a yardstick should measure the development of real welfare, welfare being understood to mean the satisfaction of wants derived from our dealings with scarce (economic) goods. In the economy what is ultimately involved is the psychical factor, satisfaction—the quality of life, to the extent that this is dependent on scarce goods. These scarce goods encompass not only the output of production, but also other factors determining welfare such as the amount of leisure, the quality of work, future security, and income distribution.

There is agreement among economists that ultimately there is little that can be said objectively about such a subjective quantity. What is feasible, however, is to be as unambiguous as possible in tracing trends in the various factors going to determine welfare. This means that separate indicators are required for the trend of production, the trend of involuntary unemployment, and so forth. It is then up to each individual to add up the indicators and determine whether he or she enjoys greater welfare, in the strict sense of the term.

It is then essential, of course, that each of the indicators reflecting the factors determining welfare provide as true as possible a picture of the factor in question: a production yardstick, a yardstick for income distribution, and so on. For a long time now, the National Product or National Income has probably served as a reasonably true indicator of production, but because of the growing loss of environmental functions (which can be defined as normal economic goods and expressed in monetary terms; see below) and the growing number of asymmetric entries, the GNP index provides ever less insight into the net increase in economic goods. For the latter point see Chapter 11 of this report.

The asymmetric entries consist of expenditure items for compensation or elimination of environmental loss, to the extent that these are financed by government or by private households. By far the greater part of the loss of scarce environmental functions is not compensated or eliminated, however. Neither is this loss booked in the GNP index, and consequently the national income figures do not give a complete picture of the quantity of goods made available each year. Information on the availability of economic goods would therefore be vastly improved if the loss of scarce environmental functions were expressed in monetary terms. By comparing the result of such a calculation with the national income figures, an impression would be gained of the net economic result of our productive and consumptive activities. This chapter describes how such a quantification could be carried out.

Even today, the increase in production as measured by national income (or Gross National Product, GNP) is called economic growth, identified with an increase in welfare and conceived as the indicator of economic success. Among other things, such a definition excludes the scarce environment from economics. Economic growth, defined in this manner, is afforded the highest priority in the economic policies of all the countries of the world. At the same time, across the world we see growth of GNP in accordance with the present pattern of production and consumption being accompanied by the destruction of the most fundamental scarce, and consequently economic, good at man's disposal, that is to say, the environment.

The gist of this report is the quest for an indicator in which extraction from—and degradation of—the environment are accounted for, in other words an indicator for a level of activity that can be sustained indefinitely. Sustainable National Income (SNI) is such an indicator. As we have stated throughout this report, a yardstick like SNI represents something of a threat to a society such as ours, oriented as it has been towards traditional growth of GNP. When an SNI index indicates that progress is perhaps not as pronounced as we have long thought, and that there may even be a decline, it cuts to the heart of the western cultural success formula (Potma, 1990).

Hueting and his colleagues at the Netherlands Central Bureau of Statistics (CBS) have made great progress in developing the theoretical framework of a method for calculating an SNI. The one aspect still to be finalized is quantification of the effect on the national income of a direct shift from environmentally burdening to environmentally benign activities. Such a shift is required if technical measures prove inadequate or too expensive for achieving the standard of sustainability. Before the work of CBS can be elaborated to create a practical instrument and a first quantitative SNI published, however, there is still much work to be done. Given the nature of the CBS approach, this is not surprising. The approach is characterized by its comprehensiveness—the calculations encompass virtually every aspect of our consumption of the environment and natural resources, to the extent that these are relevant from the angle of sustainability. This is due to the broad way in which the concept of environmental functions is formulated (see the next section), that is, as the possible uses of our physical surroundings, which comprise both renewable and nonrenewable resources. Sustainability then means that environmental functions remain available indefinitely, in whatever form (substitution is one possibility). In comparison, other approaches are partial in character. In addition, it is characteristic of the CBS approach that sustainable use of the environment is taken as the point of departure for assigning a monetary value to environmental effects.

This process of monetarization requires that a supply and a demand curve be established. The supply curve presents no problem: it consists of the progressive costs per unit of environmental burden avoided. Earlier studies (Hueting 1974, English edition, 1980) have indicated that only in exceptional cases can preferences for environmental functions—the demand curve—be fully quantified. Expenditure on compensating for a function can be considered as revealed preferences, because provisions are made to replace the function that was originally present. Financial damage can also be conceived of as revealed preferences, since it represents losses suffered as a result of the disappearance of the function. It is clear, however, that only very few losses of environmental function are compensated, and that such losses are by no means always reflected in financial damage. Often, too, there is no possibility of compensation. When, for instance, water is polluted by chemicals, compensation of the function drinking water or water for agriculture is possible to a certain degree, by purifying the intake of polluted ground or surface water. In the long run, however, it is necessary to eliminate the source of the pollution, because of the cumulative effect. As another example, loss of soil through erosion cannot be compensated. Most important of all, much of the damage caused by losses of function will occur in the future, as is exemplified by the damage caused by the loss of climate stability, by the loss of the functions of the tropical forests (gene reserve, regulator of the water flow, preventer of erosion, supplier of wood, buffer for CO_2 and heat, climate regulator, and so on) and by the disruption of ecosystems due to the extinction of species. The risks of future damage and the resultant poor prospects for the future cannot manifest themselves via the market of today. Hueting (1992) demonstrates why, particularly in these, the most important cases, techniques[1] such as willingness to pay or willingness to accept cannot yield reliable results.

For this reason, any monetary valuation of environmental functions and their loss is based on an assumption about demand. Hueting (1986, 1989) has proposed assuming that individual economic subjects have a desire to use the environment (including natural resources) sustainably. The rationale for this assumption is the fact that in recent years countless institutions throughout the world have declared themselves in favor of economic development based on sustainable use of the environment and its resources.

---

[1] EDITOR'S NOTE: These techniques (or valuation methodologies) have been discussed in the introduction to Part IV as well.

In the next section, the concept of environmental function is elaborated. In brief, environmental functions are defined as the possible uses of our material surroundings: water, soil, air, plant and animal species, and depletable resources, including energy resources—in other words, the renewable and nonrenewable resources. In these terms, sustainability involves using the environment (and resources) in such a way that the functions remain available ad infinitum.

For renewable resources, sustainability can be achieved by maintaining or restoring the regeneration capacity of the natural processes in question. For nonrenewable resources, the functions can remain available if the annual consumption of the resource does not exceed the overall amount that can be supplied indefinitely from the resource reserve, from recycling of, and substitutes for, the resource; and from improvements in efficiency of resource use, while keeping the volume of national product maximally constant. If both the production level and the relative overall development rate of recycling, substitution, and efficiency improvement remain constant, this rate must be equal to the annual consumption of the resource, expressed as a percentage of known reserves (Tinbergen and Hueting, 1992). In this case, both the annual consumption of the reserve and the remaining reserve decrease exponentially, with depletion eventually being compensated by substitutes only.

The CBS method, now, comprises the following basic steps:

1. A supply curve for environmental functions is created by drawing elimination cost curves, elimination costs being the costs of at-source measures to eliminate the burden on the environment, thus restoring and maintaining the environmental functions. For resources, this takes the form of developing substitutes, increasing efficiency, and recycling. (The supply curve is discussed in greater detail in the section "Environmental functions," below.)

2. A demand curve is defined describing society's demand for environmental functions. The basic point of departure here is the explicit worldwide desire to achieve development based on sustainable use of the environment. (The motives underlying this choice will be explained in the section "The demand curve.")

3. The supply and demand curves are combined, permitting calculation of the aggregate cost of sustainability, which can be subtracted from the traditional GNP. (This is considered in

greater detail in the section "The main elements of the method.")

To render discussion of the methodology more accessible, in the following sections some of the basic building blocks of an SNI will first be discussed individually before we proceed to the elaboration of the method as a whole. At the end of this chapter we shall draw up the balance—What do we know once we have an SNI? How can it be used? What further developments need to be initiated?

ENVIRONMENT AND ECONOMY: BUILDING BLOCKS OF THE SNI

ENVIRONMENTAL FUNCTIONS

On closer inspection, the terms *environment* and *environmental goods* are too imprecise when it comes to allocating a cost to environmental loss or to sustainability. A good starting point for further economic analysis is to be found in the concept of environmental functions, introduced by Hueting (1970) and elaborated further in *New Scarcity and Economic Growth* (Hueting, 1980; Dutch edition 1974).

The environment, our physical surroundings, provides a large number of functions, or possible uses. Once use of an environmental function is at the expense of another (or the same) function—or threatens to be so in the future—environmental functions fully satisfy the definition of economic goods: a good is scarce when in order to obtain it one must sacrifice something else that is desired. Scarce environmental functions are the most fundamental economic goods available to humans, for the simple reason that in all our actions, including production and consumption, we are dependent on the availability of these environmental functions.

This is best illustrated by an example. Water provides a variety of environmental functions. It can be used as a source of drinking water, as swimming water, cleaning water, cooling water, and so on. However, economic development has made these functions increasingly scarce: water that has been used as a dumping ground for waste cannot simply be used as drinking water or swimming water, for instance. Similarly, through ever more intensive use of space, the function space as the carrier of natural ecosystems (and others) is lost. Plant and animal species become extinct through the loss and fragmentation of their habitat. Besides a qualitative (pollution) and spatial aspect, there is also a quantitative aspect to the environmental problem. As a result of many kinds of thoughtless use of our physical surroundings, quantities of soil are lost, and with them functions that are indispensable to humanity; similarly, resources threaten to become depleted, and with them functions that are essential for the survival of civilization.

The environmental functions can start competing with each other, or with themselves: After exceeding certain thresholds, the function "water as dumping grounds for waste" is in competition with the function "water as raw material for drinking water supply"; The function "water as accommodation for species of fish" can be in competition with itself as soon as overfishing a certain species is threatening its survival. Environmental functions therefore can be viewed as normal economic goods of which there is a certain supply and for which there is a certain demand. Reasoning further along these lines, it can be said that a loss of environmental functions means that costs are incurred, irrespective (for the time being) of whether these costs are measured in monetary terms or physical units. The key question in developing a methodology for calculating a Sustainable National Income is precisely this: how can the costs of lost environmental functions, measured in physical terms, be converted into costs measured in monetary terms? In other words: what are the costs of sustainable use of environmental functions? Some environmental functions are similar in nature to consumer goods—drinking and swimming water, and fuels, for example. Others are more like the capital goods used in production processes, for example the soil as used for crop cultivation, or process and cooling water.

By introducing the concept of environmental function, the notion of economic goods is broadened, enabling environment and production to be treated within a single framework. Expressing the physical environment in monetary terms, based on sustainability, makes confrontation with the standard national income possible.

THE VALUE OF ENVIRONMENTAL FUNCTIONS

If the regeneration or self-purification capacity of the environment remains intact, sufficient space is left for the survival of plant and animal species, and the depletion of resources is balanced out by development of substitutes and increased efficiency of use, then year in, year out, the environmental functions will remain available to the same degree. In today's terminology, this would represent a sustainable situation. When the carrying capacity of the environment is exceeded as a result of economic processes and a loss of environmental functions occurs, the value of these functions increases. In macroeconomic terms, an increase in value always expresses a decrease in prosperity. Environmental functions change from being free goods, i.e. goods that are in abundant supply relative to demand, to being scarce goods. This decrease in prosperity is a result of the process that renders produced goods ever cheaper. That process is the rise in productivity, which increases the volume of goods and services produced by man. Expressed in more straightforward terms, one can purchase ever more with the same income. We then say that there is growth of the real (national) income. On a macroeconomic scale, a fall in price thus always implies an increase in prosperity.

This increase in prosperity is reflected in growth of the real national income. This is not the case for the accompanying loss of economic goods—the decrease in prosperity. We now call this loss the cost of the growth of the national income, as it is currently measured. In economic terms, this growth does not appear to be very significant, for the loss of the most fundamental economic goods—the environmental functions—is omitted from the calculation. Nonetheless, in every country in the world this growth is considered to be an indicator of economic success. It would therefore seem desirable and useful to make an estimate of an index in which the loss of environmental functions is discounted. Such an index, rooted in the assumption that people wish to act sustainably, is provided by the Sustainable National Income.

The underlying rationale is now universally familiar. The national income is the aggregate value added created in all forms of production, expressed at market value (the added value of government services being set equal to salary payments). Environmental functions are collective goods. They cannot be individually traded and thus lie outside the market mechanism. If environmental functions could be traded, a price for them would be automatically created by the market mechanism. It should be added that it follows from the above that it is not certain whether this would lead to a sustainable situation. After all, it is merely a hypothesis to assume that people wish to act sustainably.

Differences in calculating the price of the environment can always be traced back to the various assumptions made with respect to the preferences. It is a source of confusion that in most calculations no explicit indication is given of the assumption made concerning the demand for environmental functions. This is something that remains veiled. On further analysis, it is often found that virtually zero preference is assumed for sustainable use of environmental functions (see, for example, Hueting, 1991). In the CBS approach, the crucial assumption is precisely that the standard of sustainable development reflects the demand curve for environmental functions (See the section on the demand curve). An estimate of the costs of sustainable use of the environment is of vital importance for policymaking, and not only for developing an indicator such as the SNI. The cost figures to emerge can also serve as a tool for determining the rates and structure of the taxes on which an ecologization of the tax system should be based. The basic idea behind such a tax reform is that taxes should be based on a number of crucial environmental factors such as resources, wastes, and energy resources, offset by a reduction in taxes on the labor factor.

ENVIRONMENTAL FUNCTIONS: THE SUPPLY CURVE

Construction of a supply curve for environmental functions, analogous to the supply curve for market goods is a relatively straightforward, but extremely labor-intensive element of the CBS method for calculating an SNI. This can be built up from the costs of measures required to restore and maintain the various environmental functions. A curve of this kind, considered in greater detail below, can also be termed an elimination cost curve; in it, the costs of various measures for eliminating environmental damage are set out in order of increasing cost per unit of environmental damage avoided.

In practice, it is not feasible to draw a single elimination curve. Rather, several curves must be prepared, for the costs associated with (1) spatial rezoning (e.g., for recovery of the function space for natural ecosystems, required for the survival of plant and animal species); (2) long-term conservation of the functions of resources, including energy resources (development of substitutes and increases in efficiency); and (3) at-source elimination of the various biological, chemical and physical agents. An agent is taken to mean a substance or quantity of energy added to or removed from the environment by man (in whatever form) that may cause a loss of function; examples include chemicals, plants, animals, heat, and radioactivity. The main chemical agents include acidifying emissions, greenhouse gases, and mineral nutrients. An important biological agent is the removal of plants, the clearest example being deforestation, which leads to the loss of a number of functions including regulation of the water flow, prevention of erosion, and supplying wood.

The elimination cost curves enumerate all kinds of potential abatement measures. Examples include flue-gas scrubbing, burner modification, use of catalytic converters, substitution of fossil fuels by solar-based forms of energy such as wind, tides, and photovoltaic cells, terrace-building and reforestation, reservation of space for ecosystems, and creation of an ecological infrastructure. In this way an inventory is made of the measures and their respective costs. The measures—technologies and options that are already available—are costed at their current market price. In the CBS method, such measures are successively applied until the level of environmental protection deemed necessary to guarantee sustainable use of environmental functions is achieved.

THE METHODOLOGICAL PROBLEM

In taking the approach based on the supply curve, as described above, a difficult methodological problem arises, however. It is perfectly feasible that achievement of the desired degree of environmental protection will in certain cases involve implementation of extremely drastic and costly abatement measures; so drastic and costly, at any rate, that in reality society would probably not take such steps, opting rather for a different approach. Society might give preference to a shift in the structure of production and consumption, for example, if this were anticipated to involve less sacrifice than maintaining the existing structure and combating its environmental impact using the most expensive technical means. It is also feasible that insufficient improvement can be achieved by means of technical elimination measures. For some environmental problems, implementation of the entire range of technical measures might still be inadequate for achieving a situation of sustainable use of environmental functions.

Against this background, a key question regarding the method used to calculate the SNI is how to handle such nontechnical measures, i.e. volume measures. How should the effect on the national income of a shift from environmentally harmful to environmentally benign activities be estimated?

As a first approximation, an attempt has been made to remain as close as possible to economic rationality. It has thus been assumed that as the costs of abatement techniques become excessively high, the freed up labor volume previously carrying out productive work in the environmentally harmful sectors will not stay at home unemployed but will become productive in less harmful sectors. As an illustration, if cleaner and more efficient cars plus catalytic converters plus vapour recycling systems at filling stations plus cleaner car fuels are not sufficient to meet the sustainability requirements with respect to emissions and use of space, there is only one further option: less car kilometers. In such a case, though, it is unlikely that the desire for mobility will drop to zero. Instead of travelling by car, people will in all likelihood travel by public transport and by bicycle. There will thus be less of a reduction and more of a shift in the production sphere in question. The question now is; how can such a shift be expressed in monetary terms?

An attempt has been made to answer this question on the basis of an historical-economic analysis of the relationship between growth and environment. Hueting (1981) and Hueting et al. (1992) have performed an analysis of the contribution to production growth made by sectors varying in their environmental impact. An interesting picture is obtained when the National Income is broken down into the contributions made by the various sectors and an analysis made of which sectors contribute most to its growth. It is found that approximately two-thirds of the growth of the Net National Income during the period 1965-1985 is due to only one-quarter of all economic activities (expressed in terms of manpower)—those sectors where labor productivity (output per employee) has increased most. These are also—and not by coincidence—the sectors causing greatest environmental damage, during both production and consequent consumption: industry (metal, petrochemical, and chemical industries), agriculture, public utilities, and mineral extraction. Production growth, at any rate in a closed system in which total manpower also remains constant, in fact boils down to the substitution of labor by environmental use (energy, resources, emissions, wastes, etc.) with the aid of capital goods. In turn, this capital can be viewed as an accumulation of labor and environment. Stated simply, the greatest share of production growth is due to the most environmentally damaging sectors, but as a result of the laws of supply and demand, and mechanisms of income transfer, this growth spreads itself out over the entire economy. As an illustration, the volume produced by a barber is not appreciably greater than that of his colleague 40 years ago, while his real (deflated) income, his claim to a volume of goods produced, has risen by something like a factor of four over this period. Reasoning back, then, it can be concluded that a shift of a given amount of manpower from an environmentally damaging sector to a more benign one, in which productivity growth over the years has been far less pronounced or even zero, will result in a disproportionate reduction of GNP. This solution to the shift, or reallocation, problem has been criticized from various sides. The criticism has been largely accepted. The effect of a shift in national income proves to be highly dependent on the choice of reference year (the historical year that serves as a baseline for calculating real income) while at the same time there are no objective criteria for making this choice. For this reason, the magnitude of the effect cannot be calculated.

According to some critics, the effect is virtually zero. This conclusion is erroneous, however, for several reasons. In the first place, it is unlikely that the disappearance of activities whose capacity for producing goods (i.e. productivity) has soared to an incredible height during the last few decades can be compensated for by activities whereby this capacity has scarcely increased, or not risen at all. Secondly, this conclusion leads to a striking incongruity. Because the elimination cost curve is created on the basis of increasing costs per unit of avoided environmental impact, in calculating the SNI it would then suffice to achieve sustainability exclusively by means of reallocation, and the SNI would be equal to the GNP. After all, in the view of those presenting the criticism, the associated costs are zero. The effect of a shift is therefore certainly greater than zero and also not negligibly small.[2]

CHARACTER AND PREMISES OF THE SNI STUDY

In essence, the SNI exercise comprises a comparison of two equilibrium situations: the current and a sustainable situation, at a given moment in time (comparative static analysis). The transition to sustainability is seen as occurring in a millisecond, as it were, regardless of

---

[2] At his request, in 1992 Hueting had several meetings with Jan Tinbergen, who expressed great interest in an estimation of a sustainable level of economic activity. Tinbergen, who had pondered over the criticism, was also of the opinion that in a closed economy (the basic assumption of the present study is that all countries implement the proposed measures simultaneously) such a shift would necessarily have a negative impact on national income.

how long this might actually take. In other words: time is irrelevant. This has the following implications:

1. The technological state of the art remains constant. It is evident that the technology from a future year (2010, for example) cannot be introduced in the study year (1990).

2. Because time plays no part in the transition to manufacturing the means of production for implementing technical environmental provisions, there is no need to write off existing capital goods ahead of time. As a consequence, there is no capital destruction and the transition involves no extra costs.

3. The same holds for retraining the workforce. This implies that the transfer of labor for the manufacture, maintenance, and operation of technical environmental provisions is not dictated by the requisite skills. Consequently, labor can be withdrawn from each and every sector.

4. In any given economic sector, the percentage decreases in labor, capital goods, auxiliary goods, and production resulting from the transition to sustainability are equal; this holds for all sectors. This is due to the fact that, with time, each company will balance out at its own optimum size, and adaptation to a lower level of consumer good provision will be realized via a reduction in the number of companies in the sector in question. A comparative statistical analysis thus implies a production structure consisting of a linear relationship between each of the production factors, auxiliary goods and production in each sector.

The SNI is based on a number of assumptions, several of which have already been mentioned (e.g. the preference for sustainability). At this point, we make explicit five further assumptions underlying the comparative statistical model.

5. The measures are implemented simultaneously in every country in the world. It is then not possible for the Netherlands, for example, to switch to clean production of high-quality services whilst importing goods produced in an environmentally harmful fashion, as is currently the case. Obviously, this is no way to achieve a sustainable solution to worldwide problems such as the extinction of species or the greenhouse effect.

6. As far as we can presently see, if we are to adopt a simple approach there is no better way this can be operationalized

than to introduce a still stronger assumption, namely, that with regard to the effects of implementing environmental measures the economy is considered to be closed. Under this assumption, environmental measures (technical and direct shifts) do not lead to changes in import and export flows. One of the implications of this is that all the technology required for implementing environmental protection measures is produced and applied in the Netherlands. It will be clear that this assumption is unrealistic, one reason being that the Netherlands cannot do without imports of a number of raw materials. However, rejection of basic assumptions 1 and 2 implies estimating all the import and export flows associated with the environmental measures and their effects. This would be a formidable task, while at present the SNI study cannot yield more than a rough estimate.

7. Employment remains constant. In the present context this assumption is self-evident. It means that the result aimed at in the study—a comparison of the standard national income with its sustainable equivalent—is not fouled with factors that have nothing to do with the cause of nonsustainable use of the environment (it is not, in fact, difficult to include assumptions that lead to a rise or fall in employment, making the difference between the two income figures smaller or larger).

8. Both the relative preferences for market-produced goods and services (preference patterns) and the assumed absolute preference for sustainable use of scarce environmental functions[3] remain the same before and after the transition to sustainability.[4] The preferences for sustainability must at any rate antici-

---

[3] The demand for sustainable use of the environment is assumed to be completely inelastic. This rests on the implicit assumption that there are societal obstacles standing in the way of realizing the desire for sustainability. Although this key problem lies outside the scope of the SNI study, two points can be made. First, the relevance of the Prisoner's Dilemma among countries: "Our economy will be undermined if we implement measures and other countries do not follow our example". Second, the existence of erroneous information concerning relations between employment, environment, growth and the affordability of environmental conservation. See Hueting (1995).

[4] Assumptions must always be made about wants. We deem constant wants and thus constant preference patterns to be the most plausible approach. Under this assumption, changes in the composition of the package of goods are due to the fact that technological innovations in goods production differ widely from good to good. This means that, assuming constant preference patterns, the package will

pate the measures, for otherwise a transition to sustainable use of the environment is unlikely to be achieved.

9. The physical composition of the package of goods produced via the market reflects the preferences of the economic subjects, regardless of the environment.[5]

EQUALITY OF EXPENDITURE ON TECHNICAL MEASURES AND EFFECT ON NATIONAL INCOME

The achievement of sustainability implies expenditure on technical measures. In calculating the SNI, the requisite production factors can

---

contain relatively more of those goods that become more widely available as a result of rapid technological innovation and relatively fewer goods associated with a slower rate of technological change in their production. This assumption precludes the emergence of new wants. On this point Hueting (1980) has written: "Have people always wanted to fly, or did this want only arise with the invention of the aeroplane? The fact that people have dreamt of being able to fly for thousands of years (Icarus!) and the fact that a number of people over the centuries have endeavored to build an aeroplane (Leonardo da Vinci!) suggest that the former is the case." With regard to the plausibility of constant preferences for sustainability, Hueting (1987) has presented the following postulate: "Man derives part of the meaning of his existence from the company of others. These others include in any case his children and grandchildren. The prospect of a safer future is therefore a normal human need."

[5] On (the intensity of the) wants and the purposes of economic activity, Hennipman (1962) states the following: "Economic activity serves all possible kinds of ends, which are themselves meta-economic, and which economic science . . . must take as given." Ends or wants, says Hennipman, are data that are not to be judged by economic theory and that come up only in so far as their achievement or fulfillment depends on the use of scarce means. Robbins (1952) stresses that economics is concerned with the relationships of given ends and scarce means proceeding from various assumptions concerning the nature of the ultimate data (the ends). In his words, "there are no economic ends."

Without wants, there is no scarcity and consequently no economy. It therefore follows from the above that economic theory and economic research—including the SNI study—inevitably involve making at least one assumption, namely with respect to wants. In most cases, it is in fact necessary to introduce more than one assumption. Ignoring this fact in economic studies, by not making assumptions explicit, leads to what Daly and Cobb (1989) term "misplaced concreteness."

Acceptance of basic assumption 9 depends above all on the question of whether people act according to their wants. Marcuse (1964), among others, denies that this is the case. This statement cannot be "proved," and neither can its opposite, for we have no knowledge of these wants (see above). We consider consistency between human action and human wants the most plausible state of affairs, however, because of the simple observation that people would not continue to perform work that is frequently unenjoyable or dangerous if the ensuing results were constantly disappointing (Hueting, 1980b).

be withdrawn from current production without having to involve transaction costs (see above). To determine the effect of this expenditure on the national income, we would ideally need to know the extent to which production factors are withdrawn from each individual sector of the economy. The sum of the values added of the withdrawn factors could then be calculated and thus the magnitude of the decrease in production, in other words the alternative being sacrificed, i.e. the costs. However, we do not know where the factors are being withdrawn. One approach to resolving this problem is to trace the assumptions under which expenditure on measures is equal to the effect on the national income (the costs). Withdrawal of the production factors must obviously correspond with the preferences of the economic subjects. In addition, we may proceed from linear relationships between each of the factors, auxiliary goods and production volumes. This follows from the nature of the SNI exercise (see above). Because we have no knowledge of the preference patterns, an assumption must be made on this point. The most obvious approach is to assume preference patterns which, together with the linear relationships between the factors, auxiliary goods and production volumes, lead to withdrawal such that the percentage decrease in production is the same in each and every sector (this then holds equally for the production of consumer goods). This is the case if the preference ratios remain unchanged in the event of a proportional decrease in all consumer goods.[6] A proportional decrease in the output

---

[6] We thus assume homothetic utility functions; that is to say that each indifference curve can be obtained from another by multiplying the distances of all points on the curve from the origin by a certain scaling factor, i.e. all curves are of the same form. As mentioned above, this assumption precludes extremes. This can be illustrated as follows. If the real (unknown) shape of the indifference curves was to lead to a relatively large withdrawal from an economic sector with labor with relatively low value added and minor equipment with capital goods (a sector with low capital intensity), then the effect on the national income would be less than the effect calculated on the basis of proportional withdrawal of labor and capital. The assumed proportional withdrawal would then represent an overestimate of the effect. Similarly, if the labor concerned had a relatively high value added and a high degree of capital goods equipment, the effect on the national income, calculated on the basis of proportional withdrawal of labor and capital, would be an underestimate. In a modelling approach to the SNI, a study could be undertaken to establish the shape of the indifference curves given lower real incomes of the subjects (as a result of the measures). It could then be established whether this leads, on balance, to withdrawal of more labor with a relatively low or relatively high value added, compared with the proportional withdrawal presently assumed. Research on the costs of the requisite environmental technology might also indi-

of every sector precludes extremes, such as withdrawal of all the requisite production factors and auxiliary goods from one and the same sector. The same percentage of production factors and auxiliary goods are now withdrawn from each sector, with the relative share of consumer goods in the new package remaining the same as in the current situation.

To this must be added a supplementary assumption: that the production factors required for manufacturing and maintaining the technical environmental provisions are equal to the average for the economy as a whole (this follows from proportional withdrawal).

Under these assumptions an economic equilibrium will arise whereby the effect on the national income is equal to the expenditure on technical environmental measures. Obviously, price effects may occur. Consequently, a way of financing the technical measures must be assumed that does not bring about any shifts in the overall consumption pattern (or production structure). This can be achieved by introducing a levy on subjects or products, whereby all wages decrease proportionally or all prices increase proportionally, which— disregarding, for the moment, any differences in effects on distribution—at the macro level comes down to the same thing.

It goes without saying that, in estimating the expenditure on technical environmental measures, due allowance must be made for the reduced environmental burden resulting from the decreased output due to the withdrawal of production factors.

In all cases the expenditure on the measures must be recorded as intermediate deliveries, i.e. as costs, regardless of whether the measures are implemented within the polluting sectors or elsewhere and regardless of whether they impinge on production or consumption (catalytic converters for cars, for example). Entering such expenditure as final product, i.e., as revenue, would after all be asymmetrical, since the environmental loss was not written off and consequently environmental recovery may not be written up (see the first section of this chapter). The composition of the production and consumption package after implementation of the measures is unknown. This is not of importance for the SNI study, however, because the aim is to estimate the level of activity that can be achieved sustainably and not to identify

cate, for example, that what is involved is a relatively large labor volume with high value added and high capital goods equipment, i.e. higher than average, as is the case with proportional withdrawal. If this should prove to be the case, estimates on this point would have to be revised.

the exact assortment of goods this might comprise.

Until a satisfactory solution has been found for approximating the impact on the national income of a shift away from highly polluting activities towards more benign activities, the SNI study can do no more than provide a figure indicating a minimum value for the distance to be bridged to the point of sustainability in Figure 13.1 and thus a maximum value for the Sustainable National Income. Researchers at CBS in the Netherlands have the intention to investigate whether the assumptions with respect to technical measures and shifts can be validated or improved with the aid of a comprehensive economic equilibrium model.

ELABORATION OF THE SUPPLY CURVE

The preparation of the elimination curves involves a number of practical difficulties. One of these is simply the number of man-hours required to inventory a reliable and well-documented series of measures for each individual environmental problem. Not all measures are sufficiently well-known, cost estimates may vary, and the degree to which a given measure results in environmental improvements may be debatable. In addition, certain measures may overlap, synergize, or be mutually exclusive. And it should of course be borne in mind that measures designed to alleviate environmental problem A may aggravate environmental problem B. As an illustration, many end-of-the-pipe measures such as a desulphurization plant involve additional energy consumption (CO_2 emissions) and generate additional waste. Nevertheless, with blood, sweat, and tears it should be possible to solve most of these problems and structure the data in such a way that an unambiguous procedure is developed for calculating the SNI.

AN INDICATOR AS INTERMEDIATE PRODUCT

En route to an ultimate SNI, an index will be published consisting of the sum of the expenditure on technical measures and the monetarized volume of environmentally burdening activities that should shift towards environmental benignity. This index can serve as an indicator of environmental burden, alongside the various—indispensable—indicators expressed in physical units, which can never be aggregated into a single physical unit. What (relative) weights should be attached, for example, to the health effects of smog, the consequences of species extinctions for life-support functions, the risks of nuclear power (reactor accidents, security of radioactive waste storage for thousands of years) and the risks implied by climate change? This monetarized indicator has a higher statistical confidence level than the SNI, as can readily be seen, for the magnitude of the effect on the national income of a shift from environmentally burdening to (more) benign activities can probably never be established as more than an extremely rough approximation. The indicator identifies the sums associated with the implementation of technical measures and with the environmentally damaging activities that must be rejected in the process of direct reallocation for achieving the standard of sustainability.

THE DEMAND CURVE

Drawing the demand curve for environmental functions does not make matters easier. On the contrary, it proves impossible to prepare a demand curve on the basis of (the sum of) individual consumer preferences for environmental functions. This would be the most valid demand curve, analogous to the situation for normal products. Unfortunately, though, there is no marketplace where consumers can express their preferences for the products that are environmental functions in relation to other products. Measurements of such preferences, in the form of surveys for example, are always unreliable and cover no more than one or a few environmental aspects (cf. Hueting, 1989, and 1992). In some cases, part of the demand curve can be derived from the fact that when a loss of environmental functions reaches a certain point people may decide to take compensatory measures—by building a swimming pool, for example, when natural waters are lost as a result of environmental pollution. When damage, such as ozone-related harvest losses, can be assigned directly to environmental pollution, part of the demand curve can likewise be derived. In this case there are revealed preferences, preferences that come to light because of actual behaviour. Generally, though, loss of functions can be compensated for to a minor degree only, if at all, and loss of function by no means always leads to financial damage.

The intensity of preferences for the future availability of functions cannot be established. From this it follows that the level of the discount rate, when calculating long-term environmental effects, can also not be set. Using the market interest as the discount rate for calculating the present value of long-term environmental costs and benefits means that the preferences for sustainable use of the environment amount to zero, for in that case the present value of a dollar earned 100 years from now is practically nil. This is a strong supposition, the correctness of which cannot be proven. Unfortunately most cost-benefit analyses are based on this supposition (Hueting, 1991).

A STANDARD AS DEMAND CURVE

A way out has presented itself, however. If society as a whole, as a collective, voices a preference for a certain degree of environmental protection, this desire or standard can be considered as a demand curve.

As a result of the work of the U.N. Commission, WCED, chaired by Mrs. Gro Harlem Brundtland, such a standard has, de facto, come into existence, and has been internationally underwritten at the worldwide UNCED conference held in Rio de Janeiro in the summer of 1992. This standard states that societies will strive to achieve sustainable development and sustainable use of the environment and natural resources. In essence, this establishes the global demand for environmental functions—the environment is not to be burdened further than a level at which its regeneration capacity remains intact and the desired environmental functions are maintained each year anew, ad infinitum. For natural resources, regeneration takes the form of increased efficiency, recirculation and, above all, development of substitutes such as optical fiber instead of copper wire, and solar energy instead of fossil fuels. Losses of functions due to spatial fragmentation can only be eliminated by reducing or regulating activities. Only if the functions essential for human needs remain available is it possible, in the words of the Brundtland Commission, "to meet the needs of the present generation without compromising the ability of future generations to meet their own needs."

Here again, though, the above is unfortunately considerably simpler in theory than it is in practice. What is involved is establishing the level of environmental burden that does not affect regeneration capacity. But where exactly do the limits lie? It is not primarily a task of economics to pronounce on such matters, but a certain degree of scientific argumentation is required if the global desire for sustainability—expressed as it is in rough, general terms—is to be actually employed as an instrument for estimating environmental costs. For each environmental function (in practice, for each environmental problem) an elimination cost curve must be drawn and a sustainability level established. But what counts as sustainable, and what not? The environmental sciences are not sufficiently well developed on all fronts to provide an unambiguous answer to this question, and for some problems it can be queried whether this will ever be the case. What represents a sustainable situation differs from environmental problem to problem. In the case of soil erosion, for example, the situation is clear: sooner or later the soil functions will disappear if the rate of erosion exceeds the rate of soil formation. With other problems the situation is more complex, though. As an example, consider the reinforced greenhouse effect resulting from emissions of carbon dioxide and other gases. Scientists agree that this poses a risk, i.e. that there is a chance of potential negative impact. It is even likely that there is already a chance of irreversible climate change occurring. How this kind of insight is to be translated into a sustainability standard for CO_2 emissions is far from clear, however. In these and similar cases, wherever possible, an upper and lower bound will have to be established. The solution with the least risk is consistently to be taken, because a reduction of risks is in line with the concept of sustainability.

There will always be a broad margin of uncertainty within which a normative choice must at some point be made—here, and not one step further. The approach taken by CBS is to establish sustainability standards on as broad a scientific basis as possible. It is of course conceivable that the proposed sustainability standards will be contested, but if criticism is voiced, the ball should be played back to the sciences. A demand curve can thus be constructed on the basis of two hypotheses or assumptions. The first is that the standard for sustainable use of the environment is a good measure of (the sum of) individual preferences for environmental functions. This is necessary because of the impossibility of deriving these preferences directly. The second hypothesis is that if all governments state and underwrite that they strive towards sustainability, they also in fact mean that sustainable use of environmental functions is the standard, even though they certainly do not always act accordingly. Without these two hypotheses it is impossible to develop a calculus for the costs of sustainability.

THE MAIN ELEMENTS OF THE METHOD

We can now enumerate the four main steps of the method of Hueting et al. for calculating a Sustainable National Income.

First, the relevant environmental functions are inventoried. In order to be able to work with elimination costs, however, these are not expressed in terms of the environmental functions themselves, but in terms of emissions of agents causing losses of function and in terms of use of resources and space. For the year for which the SNI is being calculated these factors—emissions and resource and space use—should be available in physical units. In this context, the relevant environmental functions are those connected with forms of environmental burdens of importance from the angle of sustainability, i.e. continuity ad infinitum. This excludes noise and stench, for example, but not acidification, eutrophication, and greenhouse warming.

Second, sustainability standards must be established for the relevant agents, resources and space. Where, approximately, does the critical limit lie? How far can the environment be burdened with a given agent without the ecosystem suffering permanent change?

Third, for each function or environmental problem an elimination cost curve is prepared, as far as is necessary for bridging the gap between the environmental burden in the study year (step 1) and the sustainability standard (step 2). Three types of measures can be included in the elimination cost curve:

1. Technical measures, i.e. modifications to production processes and products. There is an enormous variety in the possible measures, ranging from end-of-the-pipe treatment to new process designs, development of substitutes for nonrenewable resources, and rescheduling the utilization of physical space. Such measures and the associated expenditure are generally well known. Under the assumptions made, the effect on the national income is equal to the expenditure on the measures.

2. Volume shifts: shifts from environmentally damaging to environmentally benign activities, to be applied when technical measures are insufficient or unfeasible, or when they would be more expensive than reallocation of activities (see the section on Environment and Economy). In all cases, the elimination cost curve should indicate the balance between the rejected and the substitute production—in other words, the effect on the national income.

3. If measures 1 and 2 result in an unacceptably low standard of living for a society, below subsistence level for example, a decrease in the size of the population constitutes the only remaining means of achieving the standard of sustainability.

The fourth step is to determine the minimum costs that would have to be incurred to bridge the gap between the environmental burden in the study year and the environmental burden deemed sustainable, following the line of the elimination cost curve. For a graphic representation, see Figure 13.1.

The supply curve in Figure 13.1 indicates with which measures and at what cost environmental functions are produced, i.e., made and kept available. The demand curve d' is society's standard for the desired volume of environmental functions or, in other words, the degree of environmental deterioration considered acceptable, regardless of the price of the measures.

The volume of environment functions that is available now—or in the study year—is B. Functions are now less available than would be

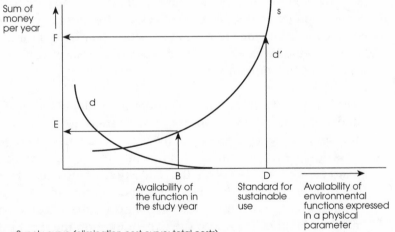

s = Supply curve (elimination cost curve; total costs)
d = Incomplete demand curve based on individual preferences (revealed by expenditure on compensation of the function, etc.)
d' = Demand curve based on the sustainability standard
BD = Distance that must be bridged in order to arrive at sustainable use of environmental functions
EF = Costs of the loss of functions, expressed in monetary terms (= effect on national income)
NOTE: The arrows indicate how losses of environmental functions recorded in physical terms are translated into monetary terms.

Figure 13.1. Translation of costs in physical units. The arrows indicate how losses of environmental functions recorded in physical terms are translated into monetary terms.

the case in a sustainable situation; in other words, the environmental burden is greater than is desirable. To move from the current to the sustainable situation, all the measures on the supply curve between B and D must be implemented. The aggregate cost of these is EF, the sustainability cost of recovering and maintaining environmental functions.

In the case of impact on a regional or global scale, national costs are calculated proportionally to the impact of the country in question.

EVALUATION: THE VALUE OF THE SNI

The above method yields an index termed Sustainable National Income. The question then is: with this index at our disposal, what does it in fact tell us?

The SNI index is primarily an indicator for the volume of goods and services that can be produced and consumed annually without exceeding a level of environmental burden that can be sustained indefinitely without jeopardizing future production and consumption options, based on the current labor volume and technological state of the art. In all cases, the elimination cost curve reflects the effect on the national income, in other words the cost of environmental loss. In a formal sense, the cost is equal to the value of the least possible sacrificed alternative.[7]

By employing this generally accepted notion of cost, the SNI can be determined by deducting from the standard national income the discrepancy (which is measured in physical terms), expressed in terms of environmental costs, between the current and a sustainable situation.[8]

---

[7] The sacrifice must therefore be both necessary and inevitable. Anything in excess of this can be termed "waste." This is known as the least means principle.

[8] Comparison of the SNI with the standard national income—the objective of the SNI study—involves the following, familiar objection, which arises with any comparison between incomes in different years. As is well known, an income figure for any single year is an insignificant number, for such figures derive their significance solely from comparison. Real national income in year $t+1$ relative to year t is obtained by expressing the income in current prices in constant prices, i.e. with the aid of a composite price index. This can only be done correctly for a constant package of goods. Because of the constantly changing composition of this package, the calculated value of the price index varies depending on the solution chosen, notably on the choice of reference year (Kuznets, 1948; Hicks, 1948; Pigou,

The SNI index derives its significance when viewed together with other economic indicators or in the context of time series. The situation does not differ essentially from an indicator such as Gross National Product. The GNP for a given year, for example, is a quantity of money that in itself provides no indication of the nature and volume of the production package this quantity represents. It is the change in this quantity that yields interesting information: is the production volume increasing or decreasing? Mutatis mutandis, similar considerations apply to the SNI. In any given year, an SNI expressed in absolute terms provides as much or as little information as a figure for GNP. The difference between the two indices is interesting, however. Much can also be learned from the trend of the SNI in the course of time, as well as from the trend of the difference between GNP and SNI.

As indicated earlier, the chosen approach leads to reproducible results that are suitable for formal processing and that also lead to a variety of useful insights. However, satisfactory solutions have not yet been found for all the problem areas involved. In particular, the question of how to deal with reallocation of (some fraction of) the labor volume from polluting to cleaner sectors still requires a great deal of thought.

1949). This unsolvable problem is certainly relevant to estimation of an SNI. For technical measures, an unchanging package of goods can be assumed (see Footnote 6, this chapter). In the case of direct shifts in activities, though, there will be changes in the package. In Hueting (1980, 1993) this objection is discussed, together with a number of other familiar objections to a comparison of national incomes for different years. An understanding of these objections inevitably leads to the conclusion that, strictly speaking, any comparison of incomes is theoretically impossible. Because national income figures derive their significance from comparison, it follows from this that the very concept of national income is impossible. This said, though, as long as national incomes continue to be calculated and compared, we too, in estimating an SNI, choose not to press the point. Once the practice of calculating and comparing national incomes is discontinued, the need for estimating an SNI largely disappears.

CHAPTER 14

Environmental Accounting and the System of National Accounts

INTRODUCTION

There has already been a long and somewhat heated debate on how national accounting should be extended towards environmental accounting (see, e.g., Ahmad, El Serafy, Lutz, 1989; Costanza, 1991; Lutz, 1993; Franz and Stahmer, 1993).

The revision of the System of National Accounts (SNA, United Nations, 1994) afforded a unique opportunity to examine how the various concepts, definitions, classifications, and tabulations of environmental and natural resource accounting can be linked to the SNA. Such linkage was originally proposed in a framework for a SNA satellite system of integrated environmental and economic accounting (Bartelmus, Stahmer, van Tongeren, 1991). Considering the knowledge currently available on environmental accounting and the divergent views on a number of conceptual and practical issues, it was not possible to reach an international consensus at that time for a fundamental change in the SNA. Nevertheless, there was agreement that the SNA should address the issue of its links to environmental concerns. Therefore, the 1993 SNA devotes a separate section (Chapter XXI, section D) to integrated environmental-economic satellite accounts and introduces refinements into the cost, capital, and valuation concepts of the central framework that deal with natural assets. This will also facilitate using the SNA as a point of departure in the development of environmental accounts.

The satellite approach to environmental accounting expands the analytical capacity of national accounts without overburdening the central framework of the SNA. The Statistical Commission, as indicated in its report on its twenty-sixth session (United Nations, 1991), endorsed the satellite approach and requested that the concepts and methods of integrated economic and environmental accounting be developed by means of satellite accounts. This approach was confirmed by the United Nations Conference on Environment and Development, which recommended in Agenda 21 that systems of integrated environmental accounting, to be established in all member states at the earliest date, should be seen as a complement to, rather than a substitute for, traditional national accounting practices for the foreseeable future (United Nations, 1992, Resolution 1, annex II, para. 8.42).

As a conceptual basis for implementing an SNA (satellite) system for integrated environmental and economic accounting (SEEA), a *Handbook of National Accounting* was published by the United Nations (United Nations, 1993).

It is not the aim of that handbook to present just another approach to environmental accounting; rather, it reflects as far as possible the different concepts and methodologies that have been discussed and applied in the past few years. The main task of the handbook is to effect a synthesis of the approaches of the different schools of thought in the fields of natural resource and environmental accounting. A thorough analysis of those approaches indicates that they are often complementary rather than mutually exclusive. This is true not only with regard to physical and monetary accounting but also with respect to valuing the economic use of the natural environment. Different analytical aims imply different valuating methods. The absence of a general approach seems to be due more to missing linkages among the different approaches than to the existence of contradictory concepts. The *Handbook* therefore does not intend to replace existing data systems like the natural resource accounts or the System of National Accounts (SNA), but rather to incorporate their elements as far as possible in order to establish a comprehensive data system.

This chapter follows that philosophy and presents concepts for integrating environmental and economic accounting which are largely based on the SEEA. Possible steps for extending SNA towards environmental accounting are described, and different valuation concepts and their consequences for deriving differing types of eco (green) domestic products are presented. In this context, the complements and differences of the approaches are addressed. Also, implementation problems are briefly raised.

BUILDING BLOCKS FOR EXTENDING NATIONAL ACCOUNTS

An SNA satellite system describing environment-related extensions of the traditional framework should have a high degree of flexibility. The needs of individual countries differ to such an extent that each country should be enabled to develop a data system suitable to analyze its own specific problems. Such differences occur—e.g., with regard to the degree of the development of the economy. Furthermore, countries with important subsoil reserves will have other problems than countries having to bear air pollution in particular. In any case, three different types of datasets could be distinguished which might allow a comprehensive analysis of the environmental-economic interrelationships:

Starting with the conventional SNA framework in monetary terms, environment-related stocks and flows could be identified by disaggregation. Such a procedure could aim at describing environmental protection activities that prevent and mitigate environmental deterioration or restore the damage (reflected in health expenditures, and material corrosion) caused by the deteriorated environment. Such an analysis of environment-related defensive activities could reveal the increasing importance of expenditures which do not aim at raising the level of welfare of the population but merely try to diminish negative impacts of a growing economy (see Leipert, 1991). In this context, environmental protection activities of the government are especially important. In the revised version of the SNA, accounts for produced natural assets as well as for nonproduced natural assets are included. Produced natural assets comprise living animals and plants cultivated by the agriculture and forestry sector. *Nonproduced natural assets* include wild animals and plants, subsoil assets, economically used areas, and water reserves. These assets are taken into account as far as they can be estimated at market values. Such accounts could be established without extending the scope of conventional national accounting and its asset boundaries.

Whereas the monetary building blocks described above are derived from the conventional SNA framework just by disaggregating monetary flows and stocks of the SNA, linked physical and monetary accounting implies extensions of the conventional national accounting system by adding physical data. Such linkages do not necessarily imply changes of the traditional SNA concepts. The additional information in physical terms could be linked with the traditional SNA framework in monetary terms without any conceptual modifications. Physical accounting could incorporate the relevant concepts and methods of natural resource accounting, material/energy balances, and input-output tabulations. Special attention could be paid to the flows of raw materials as inputs of the economic system, to changes of the economic use of land, and to the flows of residuals of economic activities discharged back into nature. As far as possible, inputs of raw materials and outputs of residuals should be linked by showing different stages of the economic transformation of raw materials which sooner or later leads to residuals which are not used anymore within the economic sphere. It seems obvious that this part might not be the most exciting one of an SNA satellite system describing environmental problems. Attempts at valuing the economic use of nature seem to be much more attractive for international discussions. Nevertheless, this part of the data system to be developed seems to be the most important one. This opinion is based on two observations: The natural environment can adequately and comprehensively be described only in physical terms, and all attempts at valuing the economic use of nature need an adequate data basis in physical units and could only provide a limited view on environmental problems.

A third type of environment-related extension of national accounting implies an imputed valuation of the economic uses of the natural environment. These approaches imply corrections of the net domestic product towards an eco domestic product. As proposed in the SEEA, different types of valuation could be applied depending on the aims of the analyses. This chapter deals in particular with the task of valuing the economic use of nature. Nevertheless, we should not forget the data basis in physical terms, which is urgently needed for such exercises. All attempts at correcting the net domestic product will be controversial. A well-elaborated data system in physical units could establish a solid basis which would facilitate judgements on different valuation concepts.

TOWARDS AN ECO DOMESTIC PRODUCT

IMPUTED ENVIRONMENTAL COSTS OF ECONOMIC ACTIVITIES

In the seventies, Nordhaus/Tobin (1973) and the Japanese Net National Welfare Measurement Committee (1973) were convinced that the gross (or net) domestic product could be modified in such a way as to make possible the measurement of the welfare connected with economic performance. They intended to purify the national product by subtracting "regrettable" parts (like national defense expenses) and added further values of nonmarket production (like household work or leisure activities) which were classified as welfare-increasing.

The optimistic belief that, with some slight modifications, economic production reflects economic welfare—has been given up. The increasing deterioration of the natural environment has brought into question the very principles of the conventional measurement of economic activities. Economic output has to comprise not only produced goods and services but also "bads" such as the by-products of economic activities like the depletion of natural resources and the degradation of land, air, and water by the residuals of economic performance.

In a similar way, inputs of economic production have to contain not only the costs of using produced assets but also the costs of using nonproduced natural assets. More important than identifying products which are still welfare-relevant seems to be the aim to achieve more complete cost accounting of economic activities. In this context, it is not intended to estimate environment-related economic welfare but to show the remaining part of the national product after deducting the costs of using the natural environment for economic purposes.

Following these arguments, an eco (green or environmentally adjusted) domestic product could be derived as follows:

Gross domestic product (GDP) − Depreciation of produced fixed assets caused by economic activities
= Net domestic product (NDP)
  − Depreciation of nonproduced natural assets caused by economic activities
= Eco domestic product (EDP).

A comparison of the net domestic product with an eco domestic product could demonstrate to what degree the results of our production activities are achieved only by destroying the natural environment. If economic activities did not cause depletion or degradation of the natural environment, the net domestic product and the eco domestic product would be identical.

The Hicksian income concept was developed at a time when there had not been any urgent worldwide and long-term environmental problems. We hesitate to apply this concept in the case of an environmentally adjusted domestic product, although others advocate it, also in this report.

In the SNA, two concepts are used for describing economic performance: the gross (or net) domestic product shows the value added produced by all institutional units resident within the country, the gross (or net) national income reflects the aggregate value of their primary income irrespective of whether it has been earned in the country or abroad. For environmental analysis, it seems preferable to refer to the geographical territory and production and household consumption within this territory. It seems to be impossible to include a description of the economic-environmental interrelationships of residents outside the geographical territory and to exclude those of nonresidents within the geographical territory (see United Nations, 1993, pp. 40–41). Thus, the SEEA uses the concept of a domestic product and that of national income.

It has been argued (see El Serafy, 1993) that the value of depreciation of non-produced natural assets should be subtracted from GDP instead of NDP. In the case of depletion, it can be argued that depleting natural resources is similar to decreasing inventory stocks which is treated as intermediate consumption. In the case of degrading natural assets, this argument seems to be not applicable because the assets are still used (with lower quality). Thus, these assets have characteristics similar to the produced fixed stocks, and their deterioration should be treated similarly to the depreciation of produced fixed assets. Furthermore, even the depletion of natural resources is normally connected with degrading (e.g., landscape) or destroying functions of natural assets (e.g., cutting a forest implies the destruction of its function as a habitat of wild animals). Thus, more arguments seem to support the treatment of natural assets as complex ecosystems which are depreciated by depletion and degradation. Of course, such an approach does not imply that the value of depreciating natural assets cannot be subtracted directly from GDP if no estimate of the depreciation of produced fixed assets is available.

PRINCIPLES OF VALUING ENVIRONMENTAL COSTS[1]

In Figure 14.1, different ways of valuing the environmental costs of economic activities are shown. There are two approaches which differ fundamentally:

1. Should the analysis focus on the state of the environment and its effects on the population in a specific country and a specific time period irrespective of the question of which economic activities have caused environmental deterioration and when (left side of Figure 14.1). In this case, the imputed environmental costs *borne* by enterprises, the government and households are estimated and deducted from the net value added of the economic units affected by environmental deterioration. Only the deterioration of the domestic natural environment is taken into account.

2. Should the analysis focus on the immediate environmental impacts of the economic activities of a specific country in a specific time period irrespective of the question of at what time and in which country those impacts will cause environmental deterioration? (Right side of Figure 14.1.) In this second case, the imputed environmental costs *caused* by economic activities

---

[1] The author of this chapter, Carsten Stahmer, wishes to thank Eberhard Seifert and Christian Leipert for their useful comments on this section.

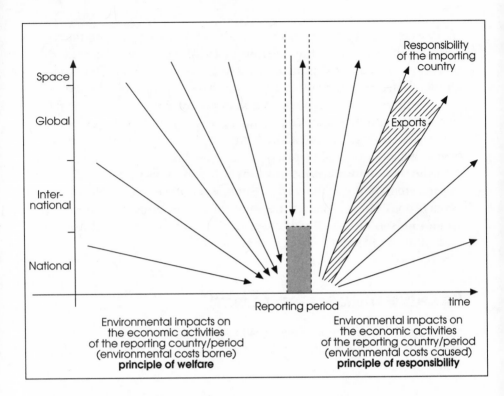

Figure 14.1 Valuation principles

are deducted from the net value added of the economic units responsible. The impacts on nature abroad are also recorded as far as they are caused by domestic economic activities. Thus, valuation refers only to domestic welfare in the first case, whereas, in the second case, the leading valuation principle is oriented towards responsibility for all countries.

The concept of environmental costs caused is based on the principle of responsibility for the long-term development of our earth. This attitude recognizes the same rights for all living beings irrespective of whether they live in our country or abroad, and is based on the ethical postulate that we should act in a way that does not adversely affect other living beings now or in the future. This principle represents a strong sustainability concept: Our economic activities should be limited to those which will not entail a decrease in the natural capital. This concept allows the substitution of one type of natural capital for another but not replacement of natural by man-made capital.

It seems to be more and more difficult to link the two types of environmental costs. Only if economic activities cause environmental deterioration in the same country as well as in the same time period are such linkages at all possible. It is typical of most of the environmental impacts of economic activities that they cause long-term and international problems. The traditional cost-benefit analysis which refers to such linkages is nowadays increasingly restricted to specific regional environmental problems (e.g. noise, changes of land use). In the SEEA, no attempt is made to link the approaches of environmental costs borne and environmental costs caused. The optimistic attempt to compare benefits and costs of economic activities in a specific country and specific time period seems to have been abandoned after observing the long-term and global problems of environmental deterioration caused by economic activities.

It seems to be very important that the environmental impacts of international trade are taken into account in calculating the eco domestic products of the importing and exporting countries. In Figure 14.1, it is indicated that the importing country has to bear responsibility for the environmental deterioration caused by the production of the imported goods in the delivering (exporting) country. A rich country could "export" its environmental problems by importing all goods whose production causes environmental problems. In a similar way, a rich country could export its dangerous wastes in order to store them in poorer countries.

In the following section, different ways of valuing environmental costs borne and caused are described. It should be stressed that these valuation methods are *complementary* and not mutually exclusive. It seems to be impossible to identify one true value of the natural environment and the changes it is subjected to. Valuation could only be oriented towards specific aims and has to differ if the analysis focusses on different aspects of environmental-economic interrelationships. The struggle between different schools of valuation seems to have only the result of wasting time and should be reduced as far as possible. We have already mentioned that it is one of the tasks of the United Nations *Handbook of National Accounting* (United Nations 1993) to support concepts which are based on a synthesis of different types of valuation. The problems of environmental deterioration are so urgent that close cooperation of all scientists working on the analysis of these problems is strongly desired.

In this context, the special task of statistical offices will be to develop the core systems of national accounts in such a way that environment-related flows and stocks can be identified. The establishment of satellite systems could be a common task of statistical offices and research institutions: Valuing the economic use of the natural environment implies modelling to a great extent. This is especially true for the concept of "costs caused" which necessitates thinking about alternative strategies of economic performance and about the societal and economic costs connected with them.

ENVIRONMENTAL COSTS BORNE

In this section two types of valuation methodologies will be discussed, market valuation and so-called "contingent valuation." This will lead us to two possible steps of adjustment of Net Domestic Product.

Market valuation: EDP I In the revised SNA (United Nations, 1994), not only flow but also asset accounts are integral parts of the overall system. These asset accounts contain produced assets whose increase is measured as gross capital formation (increase of GDP) and whose decrease is taken into account as consumption of fixed capital (decrease of NDP). The asset accounts of the revised SNA also include nonproduced natural assets (like subsoil assets, water, wild animals and plants, land) as far as these assets can be assessed at market values.

The volume changes of nonproduced natural assets are shown in the SNA in separate asset accounts and do not affect the flow accounts relevant for calculating NDP. A first approach at calculating the eco domestic product (EDP) could be to identify the volume changes (due to economic causes) of nonproduced natural assets in the SNA asset accounts and to introduce them as additional costs (depreciation of nonproduced natural assets) in the flow accounts.

Such shifts lead to the following concept of an eco domestic product:

> Net domestic product (NDP) − Depreciation of nonproduced
> natural assets at market values
> = Eco domestic product (EDP I).

The value of the depreciation of non-produced natural assets could comprise the depletion of subsoil assets, water, and nonproduced biota, or the degradation of land by soil erosion. Such market values could be derived by using observable market prices, by estimating the discounted value of expected net proceeds or—in the case of depletable natural resources—by calculating the net rent of resources (actual market price minus actual exploitation costs, including a normal rate of return of the invested produced capital). Of course, such a type of valuation only covers ecological problems as far as they are reflected by changes in economic values. Only if economic units like enterprises or the society as a whole recognize decreased market values of their assets will an eco domestic product differ from the net domestic product. The limitations of this approach, which has been supported and applied especially by the World Resources Institute (see, e.g., Repetto, 1989), are evident: Market values can only reveal pressures on natural assets and cannot reflect the whole range of economic uses of the natural environment. On the other hand, this type of valuation could be relatively easily derived from observable statistics and does not necessitate additional imputations in national accounting. The values are already recorded in the asset accounts. Furthermore, changes of market values might be the type of valuation which facilitates a discussion with politicians on the depletion and degradation of natural assets—If they understand that their country is gradually losing its natural wealth even from a narrow economic point of view, they might be willing to take into account environmental problems. The same holds true for enterprises which have to bear decreases of the market values of their nonproduced natural assets (like owned land or subsoil assets they have the right to exploit).

The proposal of Hill and Harrison for calculating the market value of depletion of natural resources is an important step towards giving specific advice on how to apply this valuation method. They argue that the value of depletion should also be taken into account in calculating a net domestic product in the core system. As long as the present SNA 1993 is applied, their proposal seems to be applicable only in a satellite system. Nevertheless, it can be argued that in the next revision of the SNA, the proposed modifications should be introduced in the core system. In the meantime, experience in compiling these values could be gathered and used for developing generally accepted international recommendations.

Another interesting approach in calculating the market value of depleting natural resources was developed by Salah El Serafy (see e.g. El Serafy, 1993, and the description in Chapter 12). His calculation method, which is a special case of the formula of Hill/Harrison, allows an analysis of income applying the concept of weak sustainability.

Such an approach seems to be especially useful in the context of the concept of costs borne which is based on a comprehensive economic point of view. That type of concept has to allow substitution between produced and nonproduced (natural) assets because it aims at maximizing income and consumption of the population of the respective country without taking into account environmental constraints. In this sense, the El Serafy approach can be recommended as a suitable concept for applying purely economic reasoning to environmental problems.

The presented concept of environmental costs borne at market values is restricted to marketed parts of the natural environment. Such an approach is especially useful for identifying environmental costs borne by enterprises because these economic units normally take into account only environmental changes reflected by the market. To a certain extent, households also suffer from changes in market values due to environmental effects, e.g., changes in the value of dwellings due to increasing traffic noise or air pollution. Nevertheless, the major part of environmental effects on human well-being (e.g. air and water pollution, noise, destruction of landscape used for touristic purposes, etc.) are not adequately reflected by changes in market values. In these cases, indirect methods for measuring the changes in these environmental services have to be developed. Such estimates could be added to the calculations at market values to achieve a more comprehensive concept of environmental costs borne.

Contingent valuation: EDP II　The most prominent valuation method for estimating the nonmarket costs borne are the different types of contingent valuation, especially the so-called "willingness-to-pay" approach (OECD, 1989). The contingent valuation method is not without controversy. In particular, it has been argued that the amount of money that people are willing to pay to improve the natural environment does not necessarily correspond to the amount that they would actually pay (free-rider problem). Furthermore, a fully developed knowledge of the quality of the natural environment and of the possible impacts on health, for instance, does not exist generally. It is therefore difficult to translate environmental effects into monetary expenses. Willingness to pay will also depend on the income situation of the individuals questioned.

Nevertheless, it seems important in a democratic and participatory approach, to take the opinions of the people into account, even if their knowledge of the natural environment is incomplete. The SEEA (United Nations, 1993, par. 320–331) proposes to ask people to what degree they would be willing to reduce their consumption level or change their consumption pattern if such actions implied an improved quality of their natural environment. In this case, the difference between the expenditures connected with existing consumption activities and those connected with the offered change in such activities could be used to represent the value of the environmental quality lost. It is like a fictitious market value of environmental quality.

The willingness to forgo consumption involves at least actual repercussion costs of households (for example, environment-related health expenditures, additional commuting and housing costs). If such a deterioration could be avoided, households would of course be willing to reduce their respective defensive expenditures. Studies on the willingness to reduce household consumption levels could therefore identify actual repercussion costs as part of the total amount of voluntary reductions and, in a second step, focus the questions on the willingness to pay for additional reductions. This approach stresses again the importance of identifying actual environmentally related defensive expenditures because actual repercussion costs are an important part of them.

When concepts of contingent valuation are applied, more comprehensive estimates of environmental costs borne lead to a second type of eco domestic product:

Net domestic product (NDP) − Depreciation of nonproduced natural assets at market values
− Depreciation of nonproduced natural assets at contingent values
= Eco domestic product (EDP II).

Of course, the estimates at contingent values have to refer only to environmental changes which have not yet been recorded at market values.

Further studies are necessary on how to implement the contingent valuation approach in an economic information system. The concept has been developed at a microeconomic level and it seems to be difficult to apply it at a macroeconomic level as well. A pragmatic procedure for national accounting could be to start with focussing on estimating actual repercussion costs as part of contingent values and leave more comprehensive estimation to further research.

ENVIRONMENTAL COSTS CAUSED

As already mentioned, the concept of environmental costs caused is based on the principle of strong sustainability. Our economic activities should be limited to those which will not entail a decrease in the natural capital.

Avoidance cost approach: EDP III The suitable valuation concept for such an attitude is the avoidance (prevention) cost approach (see especially Hueting, 1980, 1993). We measure the decrease in the level of economic activities by the additional costs necessary for achieving sustainable development and interpret this decrease as the value of the natural environment which reduces the net domestic product.

On the other hand, the effects of present economic activities comprise not only impacts in the same period but also in the future. This concept is shown on the right-hand side of Figure 14.1.

> Net domestic product (NDP) − Depreciation of nonproduced
> natural assets at avoidance (prevention) costs
> = Eco domestic product (EDP III).

It should be stressed that the necessary prevention costs are only calculated for the impacts of domestic economic activities of the reporting period. Thus, prevention costs comprise the costs of activities which prevent at least further negative impacts of present activities (in the reporting country or abroad). It could happen that the environmental quality would decrease even if no additional negative effects of present activities were added. In this case, the negative development of the natural environment in the reporting period was already recorded in the past by calculating prevention costs for the economic activities of previous periods.

The calculation of avoidance (prevention) costs is only possible on the basis of modelling. In the process of thinking about alternative and more sustainable ways to achieve economic performance, the comparison between the actual and the desirable development can only be a hypothetical one. The necessary modelling work can be more micro- or more macro-oriented.

In the first case, each economic activity is studied with regard to its environmental impacts. In a second step, alternatives are developed which avoid possible negative environmental effects of the actual activities. The difference between the net value added of the two types of activities is treated as (imputed) environmental costs. By adding the differences of all economic activities studied we obtain the total of environmental costs which is subtracted from the net domestic product. The main problem with this approach consists in the dependencies between the economic activities and their impacts on the natural environment which allow only a restricted additivity of the environmental costs calculated at the micro-level.

An alternative way of calculating avoidance costs is to introduce limits (standards) of environmental effects of economic activities first and to calculate in a macroeconomic model a sustainable level of economic activities (especially of final consumption of products). The difference between the actual net domestic product and the hypothetical net domestic product could be interpreted as the necessary environmental costs. In this case, the eco domestic product would be the net domestic product of the hypothetical economy without negative impacts on the natural environment.

A crucial point of all estimates of avoidance (prevention) costs is the determination of the level of economic activities whose effects on the natural environment could be assimilated by nature without long-term negative impacts. This determination is especially difficult because we have to take into account effects which could be both long-term and widespread. If knowledge about these effects is limited, a risk-averse attitude should be applied. Furthermore, setting sustainability standards also means solving distribution problems. If, for example, global limits for producing carbon dioxide have been set, it will have to be decided which proportion of the globally allowed pollution of carbon dioxide is accepted for the individual countries. Theoretically, the principle that pollution per head should be equal worldwide seems to be acceptable. If developing countries accept a lower level of pollution they should receive a compensation from the (normally richer) countries whose pollution per head exceeds the average. The methodological problems mentioned here have been elaborated in Chapter 13.

The strategies to avoid negative environmental impacts of economic activities differ with regard to the types of economic use of the natural environment:

1. In the case of depleting nonrenewable natural resources (like subsoil assets), the quantitative decrease in these resources could be diminished by developing more efficient ways of using

raw materials. Nevertheless, a decrease in these assets will normally be unavoidable. In this case, replacement by other types of natural capital would be necessary to achieve at least a constant level of natural capital as a whole (Daly, 1991). The substitution costs could be used as estimates of the environmental costs.

2. In the case of depleting renewable cyclical natural assets, natural growth (biota) or natural inflow (groundwater) should be balanced against the quantities depleted. If depletion exceeds natural increase, the necessary reduction in net value added of the depleting industries can be used as an estimate of environmental costs.

3. In the case of land use, sustainability implies a constant qualitative and quantitative level of landscapes and their eco systems, including their biodiversity. If an increase in economic activities results in a decrease in this level, the necessary reduction in economic activities and in their net value added involved will have to be calculated.

4. With regard to discharging residuals into nature, numerous possible prevention activities have to be analyzed which comprise the replacement of products (by increasingly more environmentally friendly goods and services), technological changes to produce technologies with low pollution, and a reduction in economic performance, especially lowering the consumption level of the population. To reach specific standards, the strategy based on minimal costs should be chosen, and these prevention costs represent the deterioration of non-produced natural assets caused by pollution etc.

The different types of economic use of the environment can be specified in terms of environmental problems. Based on these specific environmental problems a so-called elimination cost curve can be constructed. This has been elaborated earlier in Chapter 13.

Modelling connected with the avoidance cost approach implies a variety of assumptions which do not fit into the traditional national accounting system. It exceeds the traditional scope of work of statistical offices. This fact should not be interpreted as a disadvantage of this method. Urgent environmental problems necessitate a close cooperation between statisticians and model builders. Improved knowledge in environmental analyses can only be achieved if specialization is given up and interfaces of different disciplines are defined.

Alternative developments of environmental quality In Figure 14.2, alternative developments of the environmental quality in the reporting period t are shown. The actual development (2) indicates a diminished decrease in comparison to a (hypothetical) situation of no actual environmental protection activities in t (1). If the economic activities in t had no negative impacts on the natural environment, the development would be indicated by the (hypothetical) line (3). To achieve a constant level of environmental quality in t, restoration activities in t would be necessary (4) which not only balance the negative impacts of the economic activities in t (3) but also undo negative impacts of the past. It may happen that the natural environment is already destroyed to such an extent that an improvement of the environmental quality (5) is urgently needed. In this case, further restoration activities will be necessary. Of course, scenarios (4) and (5) will only be feasible if the deterioration of the natural environment is reversible. This assumption seems to be unrealistic to an increasing extent.

Additional restoration activities: EDP IV If additional restoration activities were undertaken which, for example, balance the deteriorating impacts of present economic activities (see (3) in Figure 14.2) or even prevent a decrease in environmental quality (4), or increase the quality of the natural environment (5), the costs incurred could be recorded as gross capital formation in nonproduced natural assets which balance or exceed the depreciation of natural assets caused by the impacts of economic activities in the present reporting period. It seems to be very important to show in an integrated environmental and economic accounting system not only failures but also all attempts made to improve the quality of the natural environment. This could play a central role within the environmental protection activities of the government. If restoration activities were observed, an eco domestic product could be derived in the following way:

Net domestic product (NDP) − Depreciation of nonproduced natural assets at prevention costs
+ Increase of nonproduced natural assets by restoration activities (gross capital formation)
= Eco domestic product (EDP IV).

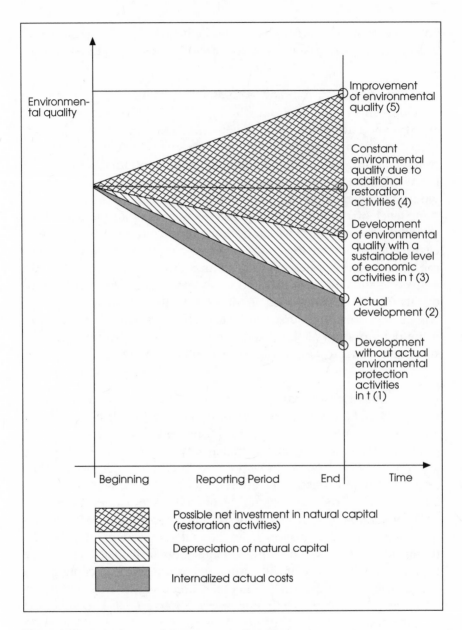

Figure 14.2 Development of environmental quality

When there is a positive net capital formation (gross capital formation minus depreciation) in nonproduced natural assets, the eco domestic product will be higher than the net domestic product.

The introduction of the concept of gross capital formation for non-produced natural assets reflects a very important step from defensive to offensive actions in the field of environmental policy. The concepts of environmental accounting should contribute to such an active attitude. If enterprises, households, or government restore or improve the natural environment, this should be reflected in the accounting system to show not only defeats but also success. The title of the last conference of the Society of Ecological Economics, "Investing in Natural Capital," reflects this attempt.

International trade: EDP V An important extension of the concept of environmental costs caused takes into account the environmental impacts of international trade. After estimating environmental costs borne in line with the method described, an input-output analysis could be carried out to calculate the direct and indirect environmental costs connected with goods and services exported. These costs should be subtracted from the environmental costs of the exporting country and added to the environmental costs of the importing country. Such corrections could be done for all internationally traded products whose production involves severe environmental problems. In this case, the corrected eco domestic product would be as follows (without taking into account restoration activities):

Net domestic product (NDP) − Depreciation of nonproduced
natural assets at avoidance costs (caused by all domestic economic
activities)
  + Depreciation of nonproduced natural assets at avoidance costs caused
    directly or indirectly by exported products
  − Depreciation of nonproduced natural assets at avoidance costs caused
    directly or indirectly by imported products
  = Eco domestic product (EDP V).

The concept of EDP V stresses the importance of international trade for environmental analysis and allows a more comprehensive concept of responsibility. Furthermore, it facilitates the necessary discussion of the extent to which environmental problems are exported from the rich developed countries to the poorer developing countries. The disadvantages of this approach are the necessary modelling work of indirect environmental costs and the substantial amount of basic

input-output data necessary for establishing international links between economies.

Concluding Remarks

Implementation of Integrated Environmental and Economic Accounts

To allow an adaptation of the SEEA to different environmental and
socioeconomic conditions in the individual countries, it has been designed to be as comprehensive, flexible, and consistent as possible.
The aim of comprehensiveness refers to both a variety of patterns of
economic development or categories of environmental deterioration,
and alternative theoretical approaches. Data availability and the potential of further improvement of the database restrict the application
of SEEA concepts. These constraints necessitate a flexible system
which should comprise a variety of building blocks which could be
used independent of each other. This necessary flexibility of the SEEA
should not affect the consistency of the system. A consistent data
system will be ensured if the versions of the SEEA remain an extension
of the national (economic) accounts and apply the accounting rules
of extended accounts.

The implementation of the SEEA should focus on high-priority concerns and related economic activities. In addition, the implementation
is still limited by data availability. As mentioned in the intoduction of
this chapter, the SEEA consist of three main building blocks. Each
building block has specific data requirements. Data collection for the
implementation of some building blocks may require data compiled
for other parts of the system.

The compilation dependencies among the different building
blocks of the SEEA indicate that physical data and accounts need to
be established first. Monetary data could then be estimated in a second
step. This procedure does not exclude the immediate implementation
of monetary building blocks that are either already available or less
dependent on physical data.

For the interested reader, more detailed information on the implementation aspects of the SEEA can be found in the appendix to this
chapter, and in the *U.N. Handbook on Integrated Economic and
Environmental Accounting.*

The foregoing material discusses the route to the implementation of the SEEA. Besides the structural, stepwise implementation, it is of great importance to perform and evaluate pilot studies like the Mexico case. Experts and policymakers can learn a great deal from these experiences. Experts, from a conceptual and empirical point of view; policymakers in terms of the possibilities and limitations of the system.

The complexity of the task of describing the environmental-economic interrelationships should not discourage the statisticians involved in this work. Basing the development of the data system on a stepwise approach with limited aims at the beginning may reduce the danger of being disappointed and overwhelmed by the difficulties. The task of supporting an integrated economic and environmental policy seems to be so important that every effort should be made to overcome the obstacles.

APPENDIX 14.1

Implementation of Integrated Environmental and Economic Accounts

 To allow an adaptation of the SEEA to different environmental and socioeconomic conditions in the individual countries, the SEEA has been designed to be as comprehensive, flexible and consistent as possible. The aim of comprehensiveness refers to both a variety of patterns of economic development or categories of environmental deterioration, and alternative theoretical approaches. Data availability and the potential of further improvement of the database restrict the application of SEEA concepts. These constraints necessitate a flexible system which should comprise a variety of building blocks which could be used independent of each other. This necessary flexibility of the SEEA should not affect the consistency

Table 14.1

PRIORITIES FOR IMPLEMENTING THE SEEA (SATELLITE SYSTEM FOR INTEGRATED ENVIRONMENTAL AND ECONOMIC ACCOUNTING)

Environmental issues	*Physical counting*		*Monetary accounting*	
	Developed country	Developing country	Developed country	Developing country

1. Use of natural assets (except discharge of residuals)

Depletion of:

1.1 Biological assets	+	+ +	+	+ +
1.2 Subsoil assets	+	+ +	+	+ +
1.3 Water	0	+ +	0	+ +

Degradation of land (landscape)

1.4 Restructuring (urbanization, changes in land use)	+ +	+ +	+	0
1.5 Agricultural use (soil erosion)	0	+ +	0	+ +
1.6 Recreational use	+	+	+	+

2. Product flow analysis	+ +	0	0	0

3. Degradation of the natural environment by discharge of residuals

3.1 Wastes and land contamination	+ +	0	+	+
3.2 Waste water	+ +	+	+	+
3.3 Air pollution	+ +	+	+	+

4. Actual environmental costs

4.1 Environment protection activities			+ +	+
4.2 Damage costs			+	0

+ +: high priority; +: medium priority; 0: low priority

of the system. A consistent data system will be ensured if the versions of the SEEA remain an extension of the national (economic) accounts and apply the accounting rules of extended accounts.

The implementation of the SEEA should focus on high-priority concerns and related economic activities. The restricted funds for statistical work do not allow a complete description of environmental-economic interrelationships. In Table 14.1, possible priorities for implementing the SEEA are shown (see United Nations, 1993, p. 153). This list divides priorities according to the type of statistical unit (data in physical or monetary terms) and according to the type of country (developing or developed).

Implementation will be further limited by data availability. Therefore, it seems useful to start by implementing those parts of the SEEA that have high priority and a sufficient data basis. After improving the database, more complete versions of the SEEA can be implemented. In Figure 14.3 (see and of chapter and see United Nations, 1993, p. 150) an overview is given of possible building blocks of the SEEA. Of course, each building block comprises a variety of specific items that have to be compiled separately (for example, accounts for different types of products, raw materials and residuals).

The arrows in Figure 14.3 represent dependencies in compiling data for the different building blocks. Data collection for the implementation of some building blocks may require data compiled for other parts of the system. For example, monetary data can, in many cases, be compiled only on the basis of sufficient physical data.

The compilation dependencies among the different parts of the SEEA indicate that physical data and accounts need to be established first. Monetary data could then be estimated in a second step. This procedure does not exclude the immediate implementation of monetary building blocks that are either already available or less dependent on physical data.

The complexity of the task of describing the environmental-economic interrelationships should not discourage the statisticians involved in this work. Basing the development of the data system on a stepwise approach with limited aims at the beginning may reduce the danger of being disappointed and overwhelmed by the difficulties. The task of supporting an integrated economic and environmental policy seems to be so important that every effort should be made to overcome the obstacles.

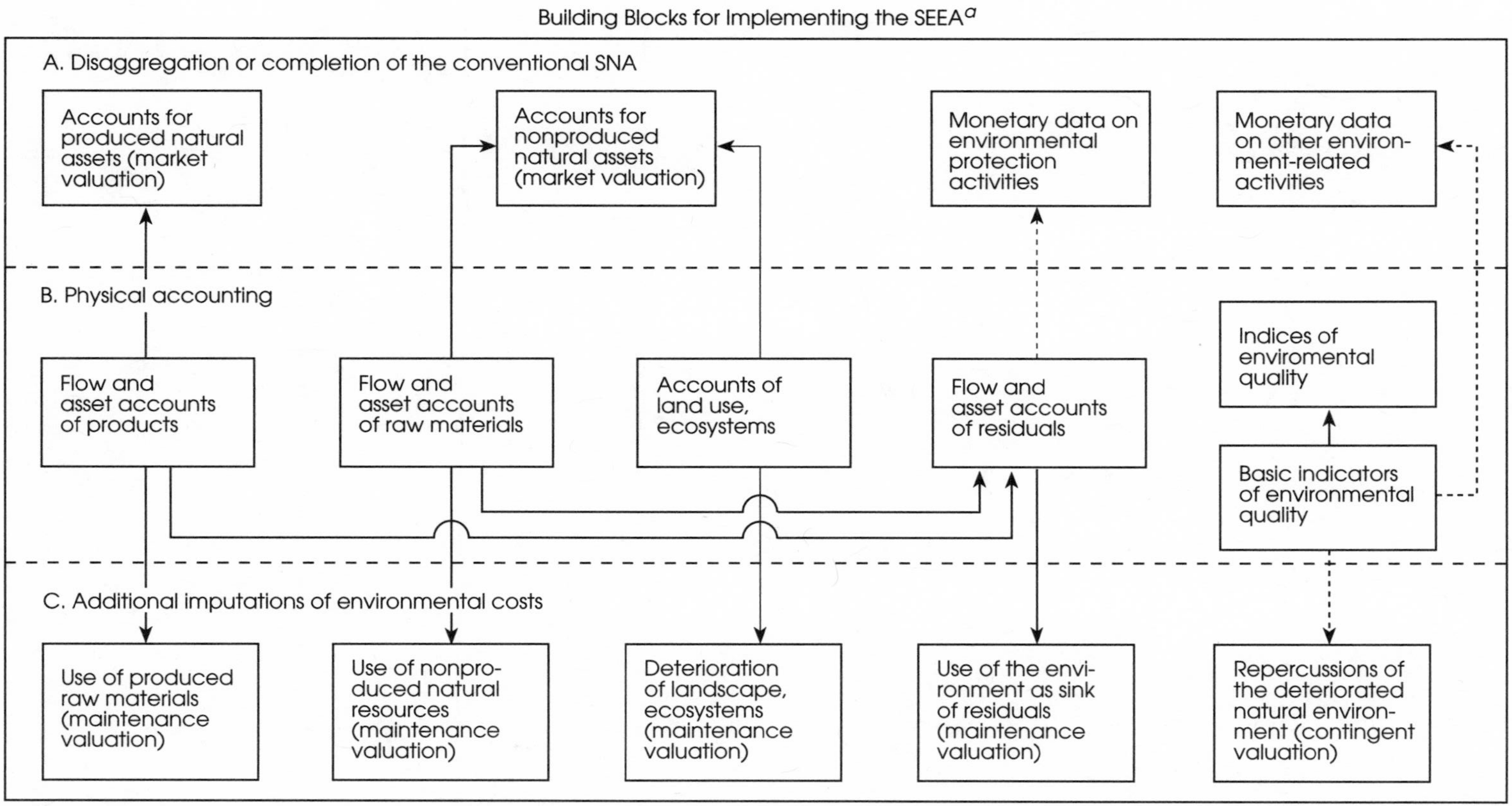

Building Blocks for Implementing the SEEA[a]

A. Disaggregation or completion of the conventional SNA

| Accounts for produced natural assets (market valuation) | | Accounts for nonproduced natural assets (market valuation) | | Monetary data on environmental protection activities | | Monetary data on other environment-related activities |

| Indices of enviromental quality | | Basic indicators of environmental quality | | Repercussions of the deteriorated natural environment (contingent valuation) |

B. Physical accounting

| Flow and asset accounts of products | | Flow and asset accounts of raw materials | | Accounts of land use, ecosystems | | Flow and asset accounts of residuals |

C. Additional imputations of environmental costs

| Use of produced raw materials (maintenance valuation) | | Use of nonproduced natural resources (maintenance valuation) | | Deterioration of landscape, ecosystems (maintenance valuation) | | Use of the environment as sink of residuals (maintenance valuation) |

[a]SEEA: Satellite System for Integrated Environmental and Economic Accounting.

Figure 14.3

P A R T · F I V E

Taking Nature into Account: Conclusions and Recommendations

1. INTRODUCTION

BACKGROUND OF THE "TAKING NATURE INTO ACCOUNT" PROJECT

This report on the economic growth, measured as GDP per capita in relation to natural resources and environmental degradation, and how this relation should be reflected in the System of National Accounts, was initiated some years ago by members of the Club of Rome (Wouter van Dieren, consulted by a.o. Martin Lees and Ernst von Weizsäcker), and has been carried out by the Institute for Environment and Systems

Analysis in the Netherlands (IMSA).[1] The project encompasses preparation of this report and research into the most opportune proposal concerning adjustment of the Gross Domestic Product (GDP) and Net Domestic Product (NDP) for environmental costs. The conclusions and recommendations presented are based on an assessment of the contents of the preceding chapters, desk research, interviews, attended meetings, and consultations with experts.

The basic point of departure is the work of Hueting and his colleagues on the calculation of the national income in a (hypothetical) environmentally sustainable economy: the Sustainable National Income (SNI), holding that the difference between the traditional national income and their estimate of SNI reflects the distance from sustainable use of environmental functions in monetary terms. As an editor, I soon experienced the breadth and depth of the conceptual and methodological discussions between conventional and environmental economists, national accountants and statisticians, and the ecologists that are neither hindered nor biased by economic or statistical training.

The diversity of views and issues in this field, which are sketched below, is enormous. We will provide a birdseye view.

There is a relatively old tradition of endeavouring to change GDP into a single yardstick, a single index reflecting welfare or progress. This index approach is strongly criticized by those who support the development of a set of disaggregated welfare indicators. Others promote methodologies aiming at an adjusted NDP reflecting Hicksian income, which is criticized by, inter alia, national accountants who put forward that the Hicksian income concept is not meant to be—and never will be—related to the System of National Accounts (SNA). The majority of national statistical offices are reluctant to relate (monetary) values to the SNA in general and GDP in particular,[2] and opt for environmental accounting in physical terms. Some statistical offices have developed their own systems of natural resource or environmental accounting. In the meantime, the U.N. Commission on Sustainable Development is working on economic, social and environmental indicators (Sustainable Development Indicators, or SDIs) which will enable us to assess the progress toward, and the distance from, sustainable development. The System of National Accounts was revised in

---

[1] IMSA is the Dutch abbreviation for Instituut voor Milieu-en Systeemanalyse.
[2] See Section 8, "Resistance to and Solutions for Change."

1993, and the U.N. has published a *Handbook on Integrated Environmental and Economic Accounting.* Parallel to the development of the *Handbook,* the methodologies it proposes have been tested by carrying out two case studies, in Mexico and Papua New Guinea.

From the above it can be seen that the scope of this report has been extended far beyond its original point of departure—the SNI according to Hueting—which will be placed in its proper context in the following paragraphs. This book has become a collection of the views of experts, who have elaborated all the themes in the various chapters of this report,[3] but with the editorial hand in the end as the final responsible.

Based on this exploration, I have drawn conclusions and formulated recommendations, which focus primarily on the role of GDP, the alternatives for measuring progress, the adjustment of GDP, and NDP, resistance to change, and strengthening international coordination. The final section, our conclusions and recommendations are summarized.

GDP: STEERING BY THE WRONG COMPASS

GDP is an important indicator for economic growth, which is associated with a high(er) standard of living and increased welfare of a nation's citizens. This is not surprising, because in Western countries from 1950 to about 1970, an increase of GDP led to an increase in welfare in the broadest sense.

In the same period, the System of National Accounts, originating from 1947, was refined in order to reflect changes in the economy and respond to changing analytical and policy concerns. This System of National Accounts is, unquestionably, a highly advanced instrument for economic analysis and decisionmaking. Economic success and development of a sophisticated system to understand and direct the economy went hand in hand.

In the late sixties and early seventies, however, the level and pattern of our economic activities began to have a manifest impact on natural resources and the environment. These so-called external ef-

---

[3] For a more comprehensive overview of the SNA, the U.N. *Handbook,* conceptual and methodological issues, the role and visions of international agencies, see Sheng (1995).

fects,[4] which were presumed to be marginal, have meanwhile assumed significant proportions. As we elaborated in Chapter 5, by not taking natural capital into account, GDP no longer reflects economic reality. Although most economists and statisticians and some politicians are aware of these failures, statistical offices still publish annual and quarterly GDP reports without mentioning the fundamental failures of the indicator. The media, too, simply interpret an increase of GDP as economic growth, which the general public in turn associates with economic success, general progress, and even an increase of welfare. For politicians it is of course very attractive to spread this message to curry the favor of the electorate. An increase of GDP is associated with successful economic policy.[5] As Tinbergen and Hueting have stated: Society is steering by the wrong compass.

Recommendation 1

It is necessary to develop more comprehensive measures of societal progress and to improve GDP and NDP as indicators for economic progress. ▪

2. STEERING BY THE RIGHT COMPASS

To correct for the dominant role played by the traditional GDP, we favor the development, implementation, and yearly publication of a set of economic, social, and environmental indicators. This set of indicators will provide us with a more comprehensive picture of societal development and will enable us to assess societal progress more accurately.

We propose to make a distinction between descriptive indicators, reflecting actual developments (taking a picture), and comparative (or normative) indicators, reflecting developments towards sustainability (comparing the picture of today's society with an imaginary economically, socially, and environmentally sustainable society); See Figure 1.

---

[4] The negative effects of economic activities have always been largely excluded in the measurement of economic growth. They are called external effects because they are not reflected in the market conditions: they are external to the market.
[5] Some of the reasons behind this mechanism have already been elucidated in "Metaphysics of growth" and will also be dealt with in Section 8, "Resistance to and Solutions for Change."

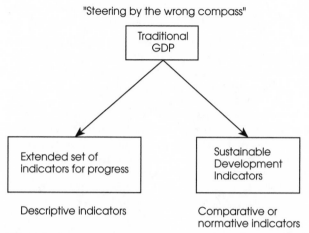

Figure 1   Descriptive and comparative indicators of progress

DESCRIPTIVE INDICATORS OF PROGRESS

Here, we can distinguish two different types of approaches: those aiming at developing a framework of indicators in which the various components of progress are shown separately (basket of indicators) and those aiming at the construction of a single index.

A more comprehensive set of indicators could, for instance, include the improved GDP and NDP, the rate of employment and unemployment, income distribution, natural resources and environmental quality, health, education, and the social security system.

Examples of a basket of indicators include the concept of Country Future Indicators (CFIs), as elaborated and promoted by Hazel Henderson and others (see Chapter 10). Examples of a single index by monetary valuation and/or aggregation of several welfare components include the Index of Sustainable Economic Welfare[6] and the Human Development Index, which have been elaborated and discussed in Chapter 10.

Although the terminology differs (indicators of quality of life, welfare, real wealth, progress, development), the basic message of those putting forward such ideas is the same: Society needs a more comprehensive set of indicators to assess societal progress.

---

[6] In late 1994, the name of the index was changed to Genuine Progress Indicator (GPI).

Recommendation 2

A wide variety of methods have been proposed to allow society to make a more comprehensive assessment of actual societal developments. As yet, most of these proposals have been developed independently of one another. It would be worth the effort to establish an international working group to make an inventory and critical assessment of the available approaches, making explicit their similarities and differences, and respective strengths and weaknesses. The next step could be to develop a more generally accepted framework, for both the basket approach (set of indicators) and the index approach. Despite the present absence of such a framework, data are available on most of the welfare components mentioned above. We therefore call on national statistical offices, politicians, and the media not to publish GDP in isolation from, but in the context of, other economic, social and environmental indicators. ∎

SUSTAINABLE DEVELOPMENT INDICATORS

At the 1992 UNCED Conference in Rio de Janeiro, the key role and the limitations of economic indices as measuring rods for progress were fully recognised. Moreover, a redefinition of progress was proposed by introducing the concept of Sustainable Development. This resulted in an urgent plea, in the Agenda 21 Action Plan, to develop so-called Sustainable Development Indicators (SDIs), since commonly used indicators do not provide an adequate indication of sustainability. Indicators of sustainable development need to be developed to provide a solid basis for decisionmaking at all levels.

To review further progress in implementing Agenda 21 and to make policy recommendations to governments and international institutions, the U.N. Commission for Sustainable Development (CSD) was established. This commission is coordinating the development of a general and consistent framework for sustainable development indicators.

Since most governments have committed themselves to using the concept of sustainable development as the compass for policy issues, there is a widespread recognition of the crucial role SDIs will play in measuring the success of future policies.

The main difference between the descriptive indicators of progress and the Sustainable Development Indicators is the introduction of economic, social, and environmental sustainability standards. SDIs will enable us to assess the distance separating the present situation of a country or region from sustainable development. The standards thus serve as a reference point against which to evaluate the present situation. In this sense SDIs are more oriented towards action and the future than the descriptive indicators. The development of SDIs implies the development of an integrated framework of descriptive statistics and standards based on the concept of sustainable development.

These standards will be based on an equal distribution of the global commons (intragenerational equity) and the needs of future generations (intergenerational equity).

As mentioned above, sustainable development includes economic, social, and environmental aspects. This is illustrated in Figure 2, which is derived from Chapter 8 of this report.

Each type of indicator is expressed in its own dimension: economic indicators in monetary terms, social indicators in human-related terms, and environmental indicators in physical terms.

At the moment, both the scientific and political community emphasize the need to develop environmental indicators for Sustainable Development. Given the urgency of worldwide environmental problems and the lack of a consistent framework for assessing environmental pressures in relation to environmental quality, this priority-setting is clearly defensible. After all, socio-economic development is constrained by environmental limitations.

Recommendation 3

Agenda 21 stresses the need for developing a consistent framework of economic, social and environmental indicators for Sustainable Development. These indicators will enable governments and supra-national organizations to assess the extent to which societies are on a sustainable path. The establishment of standards should be based on intra- and intergenerational equity. ∎

3. SNA 1993 REVISION AND THE UN *HANDBOOK ON INTEGRATED ENVIRONMENTAL AND ECONOMIC ACCOUNTING*

In its twenty-sixth session, the U.N. Statistical Commission endorsed the satellite approach for developing concepts and methods of integrated environmental and economic accounting. As a result, the 1993 SNA devotes a separate section to integrated environmental-economic

Economic objectives:

– Growth
– Equity
– Efficiency

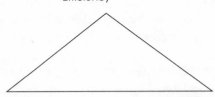

Social objectives:

– Empowerment
– Participation
– Social mobility
– Social cohesion
– Cultural identity
– Institutional
  development

Ecological objectives:

– Ecosystem integrity
– Carrying capacity
– Biodiversity
– Global issues

Social Sustainability ("SS")

ES needs SS—the social scaffolding of organizations that empower self-control and self-policing in peoples' management of natural resources (see Cernea, 1993). Resources should be used in ways which increase equity and social justice, while reducing social disruptions. SS will emphasize qualitative improvement over quantitative growth; and cradle-to-grave pricing to cover full costs, especially social. SS will be achieved only by strong and systematic community participation or civil society (Putnam 1993 a, b). Social cohesion, cultural identity, institutions, love, commonly accepted standards of honesty, laws, discipline, etc., constitute the part of social capital that is least subject to measurement, but probably most important for SS. This "moral capital," as some have called it, requires maintenance and replenishment by the religious and cultural life of the community. Without this care it will depreciate as surely as will physical capital.

Economic Sustainability ("EcS")

The widely accepted definition of economic sustainability is "maintenance of capital," or keeping capital intact, and has been used by accountants since the Middle Ages to enable merchant traders to know how much of their sales receipts they and their families could consume. Thus the modern definition of income (Hicks 1946) is already sustainable. But of the four forms of capital (human-made, natural, social, and human) economists have scarcely at all been concerned with natural capital (e.g., intact forests, healthy air) because until relatively recently it had not been scarce. Also, economists prefer to value things in money terms, so we are having major problems valuing natural capital, intangible, intergenerational, and especially common access resources, such as air etc. In addition, environmental costs used to be "externalized," but are now starting to be internalized through sound environmental policies and valuation techniques. Because people and irreversibles are at stake, economics has to use anticipation and the precautionary principle routinely, and should err on the side of caution in the face of uncertainty and risk. Human capital (investments in education, health, and nutrition of individuals) is now accepted in the economic lifestyle (WDR 1990, 1991, 1995), but social capital, as used in SS, is not adequately addressed.

Environmental Sustainability ("ES")

Although environmental sustainability is needed by humans and originated because of social concerns, ES itself seeks to improve human welfare by protecting the sources of raw materials used for human needs and ensuring that the sinks for human wastes are not exceeded, in order to prevent harm to humans. Humanity must learn to live within the limitations of the physical environment, both as a provider of inputs ("sources") and as a "sink" for wastes (Serageldin, 1993a). This translates into holding waste emissions within the assimilative capacity of the environment without impairing it and by keeping harvest rates of renewables within regeneration rates. Quasi-ES can be approached for non-renewables by holding depletion rates equal to the rate at which renewable substitutes can be created (El Serafy, 1991).

Figure 2 Comparison of social, economic, and environmental sustainability. (From Serageldin, 1993a, b.)

satellite accounts and introduces refinements to the cost, capital, and valuation concepts of the central framework dealing with natural assets.

A first step in this direction has been taken in the 1993 revision of the SNA—the asset accounts now also include some nonproduced natural assets like subsoil assets, water, and land, to the extent that these assets can be assessed at market values. In paragraphs 5–7 the degree to which the revised SNA reflects changes in natural assets will be assessed.

Meanwhile, the U.N. *Handbook on Integrated Environmental and Economic Accounting* has been developed. It was published by the United Nations in 1993. The U.N. *Handbook* serves as a conceptual basis for implementing the SNA satellite system for integrated environmental and economic accounting. It presents an overview of the various concepts and methodologies of environmental accounting that have been discussed and applied during the past few years. The main task of the *Handbook* is to effect a synthesis of the approaches of the different schools of thought in the field of natural resource and environmental accounting.[7]

Using the conventional SNA framework as a starting point, the *Handbook* presents the three following extensions of this framework:

1. Identification of environment-related stocks and flows (in monetary terms) by disaggregation of the conventional SNA framework.

2. Linking physical and monetary accounting by adding physical data. Physical accounting incorporates the relevant methods of natural resource accounting, material/energy balances and input-output data on flows of raw materials as inputs to the economic system, changes of the economic use of land, and flows of the waste products of economic activities discharged back into nature. This extension does not imply changes to the traditional SNA concepts. It is a very important extension, however, for the natural environment can be adequately and comprehensively described in physical terms only. In addition, all attempts to place a value on the economic use of nature require an adequate database in physical units.

---

[7] See also the remarks on the integration of economic and environmental accounting at the end of Chapter 10.

3. Determination of imputed environmental costs of the economic uses of the natural environment. This approach implies an adjustment of the net domestic product in the direction of an environmentally adjusted domestic product.

Recommendation 4

The U.N. *Handbook on Integrated Environmental and Economic Accounting* should be implemented by the national statistical offices as soon as possible. This process should be stimulated more intensively by UNSTAT, Worldbank, and other international agencies involved in the SNA (European Union, OECD, IMF). ▪

4. ADJUSTMENT OF GDP: DIFFERENT OBJECTIVES

In the preceding paragraphs, we have seen that GDP is not an accurate measure of welfare; that it is necessary to develop a set of descriptive indicators to assess actual development and a set of future-oriented indicators to assess the extent to which current development leads to sustainable development; and that (an improved) GDP/NDP is only one of the economic indicators of this set. In the next three sections, we focus on the improvement of GDP and NDP themselves.

Roughly speaking, two fundamentally different objectives of adjusting GDP can be distinguished (cf. Figure 3). The first objective is to adjust GDP in such a way that the modified figure reflects welfare or progress more accurately. Attempts to do so, such as the development of the ISEW or GPI, have already been dealt with in Section 2.

Another objective is to improve GDP and NDP as measures of economic performance or value added. In the next section, "The current SNA: a narrow economic point of view," the extent to which GDP and NDP reflect the costs of factor input (human-made, human, and natural capital) and the environmental costs due to economic activity will be investigated. In previous sections of this report we have seen that the current System of National Accounts does not take into account the depletion and degradation of natural assets. In this sense the current SNA represents a narrow point of view. In Section 6, "Cost accounting from a comprehensive economic point of view," the first logical and possible step to deal with this problem will be elucidated: environmental cost accounting by applying the market valuation approach for the depletion and degradation of nonproduced natural assets. In Section 7, "Cost Accounting from a Responsibility

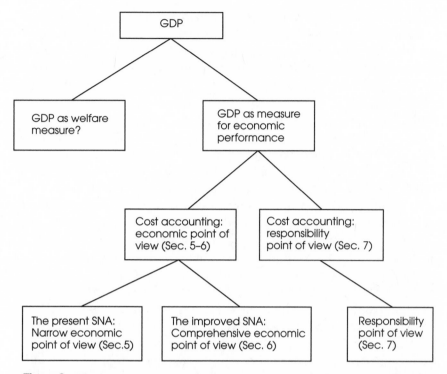

Figure 3 The line of reasoning in relation to the structure of Sections 5-7 of this chapter

Point of View," we will see that the market valuation approach is only a first step toward more comprehensive cost accounting. The next step is to estimate environmental costs from a responsibility point of view.

5. THE CURRENT SNA: A NARROW ECONOMIC POINT OF VIEW

The production account of the System of National Accounts records economic output, intermediate consumption, consumption of fixed capital, and value added, at both a sectoral and a national level. These issues have been elaborated in Chapter 2 of Part I of this report.

Gross value added is defined as the value of gross output less the value of intermediate consumption. The sum of gross value added is the gross domestic product (GDP), which is the monetary value of goods and services produced for final consumption. Net valued added is defined as the value of gross output less the values of both intermediate consumption and the consumption of fixed capital. Examples of fixed capital are machines and buildings, which depreciate through economic use. The depreciation (value changes due to economic use of productive capital) of the produced fixed assets represents a cost item, which is deducted from GDP to arrive at net domestic product (NDP). What remains is considered to be net value added.

In the 1993 revision of the SNA, both produced and nonproduced natural assets, which are also a form of productive capital, are recognised as capital assets. Value changes are now recorded in the balance sheet, as far as they can be assessed at market values.

Crops, production forests, and cattle for slaughter are examples of produced natural assets. Both value changes and consumption of produced natural assets are dealt with in the right manner, namely as changes in stocks and intermediate consumption, respectively.

Land, tropical forests, and subsoil assets like oil, gas, and minerals are examples of non-produced natural assets. According to the 1993 SNA, the value changes due to the depletion and degradation of these assets will be recorded in the balance sheet at market values.[8] Just like the value changes of fixed produced capital, the value changes of non-produced natural assets represent a cost item which should be, but is neither, deducted from GDP nor from NDP[9] in the flow accounts of the current SNA. GDP and NDP do not reflect value added, for they do not take into account the true costs of production. This represents a narrow economic point of view.

6. COST ACCOUNTING FROM A COMPREHENSIVE ECONOMIC POINT OF VIEW

As discussed in Section 5, in the SNA 1993 the value changes of non-produced natural assets will already be reflected in the balance sheet, as far as they can be assessed at market values.

---

[8] In the U.N. *Handbook on Integrated Environmental and Economic Accounting,* a more comprehensive description, both in physical and in monetary units, is proposed.

[9] Deduction from GDP or NDP depends on whether non-produced natural assets should be treated as inventories or assets (El Serafy).

Examples of value changes of nonproduced natural assets due to economic activities include the following:

- Depletion of nonrenewables: subsoil assets like oil, minerals, and groundwater reserves.
- Depletion of renewables: timber, fish, wild biota.
- Degradation of environmental media by pollution: air, water, and soil.
- Degradation of land by changes of land use.

What is necessary is that the value changes of these assets should be reflected in the flow accounts as well. In comparison with the current SNA, deducting the costs of the depreciation of these assets represents a more comprehensive economic point of view. In this approach, the value changes due to qualitative and quantitative changes in these assets during the accounting period within the domestic country should serve as a basis.

The proper valuation concept from a comprehensive economic point of view is the market valuation approach,[10] which can be applied to depletion problems and, to a somewhat lesser extent, to degradation problems. Most proposals based on market valuation deal with the depletion of natural resources. The costs of degradation problems can be included as far as the degradation of natural assets influences their market value. In the case of changes in land quality due to economic use of land, differences in market prices can be used to estimate the value of qualitative changes in areas of land.

The criterion for taking changes of natural assets into account in the flow accounts is that one can observe real market values in the economy. As a consequence, the innovations proposed in this section are consistent with "positive or descriptive economics" and the concept of value added.

To calculate the true net (environmentally adjusted) value added, the depreciation of nonproduced natural assets should be excluded from GDP. Excluded could mean: included in the intermediate inputs and, as a consequence, not included in the calculation of GDP; or

---

[10] In contrast with the approach proposed by Hueting, for instance, the market valuation approach is not based on a preference for achieving sustainability. In this context, it is used as a tool to calculate the costs of observed value changes of nonproduced natural assets.

deducted from GDP as a depreciation factor to arrive at an improved NDP.[11] For the remaining net value added this makes no difference.

For the market valuation, the most relevant proposals are the net price method and the user cost method, which have been elaborated by Repetto and El Serafy, respectively.

NET PRICE METHOD

In the net price method, the value changes of nonproduced natural assets are treated as depreciation. The depreciation factor should be deducted from NDP. This depreciation factor can be estimated by multiplying the physical change in the stock of a natural asset by the net price. Net price is defined as the market price minus the cost of discovery, extraction and marketing.

An advantage of the method is that it is relatively easy to apply. Only data on physical changes, the aforementioned costs and observable market prices are needed.

The method has been criticized for leaving GDP unadjusted. As a consequence, GDP still contains proceeds from asset liquidation which do not represent value added.

USER COST APPROACH

Starting from GDP, El Serafy proposes to make a distinction between a value added element and a user cost element as a reflection of the depletion of natural resources. The user cost element should be invested to achieve a constant flow of income in the future. To estimate true income (the element of value added) and the user cost, data on the life expectancy of the resource at the current extraction rate and a discount rate are needed. The discount rate represents the prospective returns expected from investing the equivalent of the user cost so as to yield future income that can be sustained even after the resource has been totally exhausted.[12]

---

[11] Inclusion as intermediate inputs or deduction as a depreciation factor depends on whether nonproduced natural assets are dealt with as inventories or assets.
[12] Quote from El Serafy, Chapter 12.

BATTLE OF THE SCHOOLS

Among the experts, there is no agreement yet on whether environmental defensive expenditures should be deducted from GDP/NDP and which (market) valuation methodology for depletion and degradation problems is most suitable. Unfortunately, most experts, as bearers of this knowledge, seem to be spending more time on defending their own approach than on cooperation. They elaborate the advantages and conceptual logic of their proposals and the disadvantages and conceptual flaws, from their point of view, of others.

Of course, the so-called battle of the schools is a well-known and very useful phenomenon in science because it triggers and examines innovative thoughts and ideas. With the subject at hand, however, the time has come to move beyond the scientific debate, for the following reasons:

1. the availability of promising market-based valuation methodologies;

2. the possibility of showing similarities and differences, strengths and weaknesses, and possibilities and limitations in relation to application of the methods within the SNA framework;

3. the urgency of the matter: society can no longer afford to keep steering with the wrong compass.

Recommendation 5

From a comprehensive economic point of view, net (environmentally adjusted) value added = gross output at market values minus intermediate inputs minus the depreciation of produced fixed assets minus the depreciation[13] of nonproduced natural assets (the costs of economic input in the domestic country during the accounting period). The depreciation of nonproduced natural assets must be excluded from GDP. This can be done by deducting the value changes as a depreciation factor or by including the consumption of these assets as intermediate inputs. To reflect the depreciation of nonproduced natural assets at market values in the flow accounts is a logical and necessary first step in improving the SNA and should be implemented as soon as possible. The new NDP could be calculated as follows:

---

[13] Depreciation: depletion and degradation.

Old GDP ÷ costs of depleting inventories (user cost approach)
= new GDP
÷ costs of depreciation of manmade capital (net present value method)
÷ costs of depreciation of natural assets (net price or user cost method)
÷ costs of degradation (replacement cost method)
= new NDP (EDP = NDP in transition).

Among the experts, there is no agreement on the proper market valuation methodology. We call on the experts in this area to reach consensus on the proper market valuation methodology, which can then be endorsed by the relevant international agencies involved in the revision of the SNA (UNSTAT, World Bank, IMF, EU, OECD). The statistical offices can then experiment and gain experience by first introducing this recommendation in satellite accounts. In the next SNA revision, these value changes of nonproduced natural assets should be reflected in the Core System as well. ■

COSTS OF VALUE CHANGES OF NONMARKETED NONPRODUCED NATURAL ASSETS

Because many nonproduced natural assets are not marketed, from the comprehensive (market) economic point of view, not all physically observable qualitative and quantitative changes in natural assets occurring during the accounting period within the domestic country are taken into account: they do not appear on the balance sheet nor in the flow accounts of the SNA. The costs of degradation of natural assets, in particular, remain largely hidden. The costs estimated by means of the pure market valuation approach are an underestimation of the real costs of the value changes of nonproduced natural assets due to economic activity.

One way of dealing with this problem is to estimate the costs by indirect methods. One such method is to estimate the costs of restoring or avoiding the physical observable changes within the domestic country during the accounting period. The costs of restoring the asset or of measures which could have avoided the changes in asset could serve as a monetary reflection of actual physical changes in natural assets. The estimate of the costs of these measures could itself be based on the market values of the restoration and avoidance measures. Examples of this approach are the costs of fertilizer to restore eroded soil (WRI, Costa Rica study), and the cost of re-injecting water into groundwater reserves (UNSTAT case studies).

Recommendation 6

From a comprehensive economic point of view, physical changes in natural assets during the accounting period within the domestic country which have no market value are not taken into account. Nevertheless, as these asset changes represent an additional cost item, they should be excluded from value added. ■

ENVIRONMENTAL DEFENSIVE EXPENDITURES

In the previous sections, we have dealt with the value changes of non-produced natural assets which represent an additional cost item that should be deducted from GDP or NDP.

In this section, we will deal with expenditures that have actually occurred during the accounting period: defensive expenditures. This theme has been elaborated in Chapter 11. Leipert defines defensive expenditures as expenditures that "comprise those economic activities that defend ourselves against the unwanted side-effects (negative external effects) of our aggregate production and consumption. They are understood as expenditures to cure and neutralize, to eliminate, to avoid and to anticipate burdens and damages the economic process caused, or causes, to the environment." (See Chapter 11.)

Leipert posits that those environmental defensive expenditures (translated into costs) made by governments and households which are entered into GDP are erroneously considered as value added. They should be excluded from GDP.[14] This view is shared by other experts like Hueting and Ekins. Others, like Stahmer, argue that such modifications would require concepts of economic welfare which do not appear to be applicable in national accounting. For a decision on which expenditures should be deducted, there would have to be agreement on which expenditures can be called economic output contributing to welfare, and which should be considered "economic bads" not contributing to welfare.

The U.N. *Handbook on Integrated Environmental and Economic Accounting* stresses the urgency of showing (in absolute figures or as a fraction of GDP), but not deducting from GDP, environmental defensive expenditures.

Besides the conceptual debate concerning the relationship between environmental defensive expenditures and GDP/NDP in general, there also exist several methodological problems. It is, for in-

---

[14] Excluded could mean deducted from GDP as it is presently calculated or treating these categories of defensive expenditures as intermediate consumption so they will not be reflected in GDP in the first place.

stance, still unclear how the expenditures can be translated into imputed costs for the accounting period (periodization problem) and how these actual costs are related to the additional cost items discussed previously.

Recommendation 7

Given the state of the art at the conceptual and methodological level, the choice made in the U.N. *Handbook,* namely to show the environmental defensive expenditures as a ratio, rather than to deduct them, can be defended. However, before making a final decision on how to deal with these expenditures in relation to GDP/NDP, the issues mentioned here need clarification by arranging expert meetings on them. ■

7. COST ACCOUNTING FROM A RESPONSIBILITY POINT OF VIEW

From a comprehensive economic point of view, as described in Section 6, we act as pure economists and accountants. We observe changes in our domestic assets and try to reflect them accurately in both the asset accounts and the flow accounts of the SNA. We take into account the depreciation of natural assets as far as they can be assessed at market values. As discussed in the previous section, this is only a fraction of the actual depreciation of natural assets.

From a comprehensive economic point of view, we do not take into account the impact of our domestic economic activities on other countries, on the global ecosystem and on future generations. In this sense market-based valuation represents an anthropocentric, and in a way egoistic, point of view.

Examples of the impact of domestic economic activities on the natural assets of other countries are transboundary degradation problems like the pollution of rivers or acid rain, and transboundary depletion problems like the cutting of tropical forests or unsustainable fishery. Some examples of the impact of domestic economic activities on the global ecosystem are: ozone layer depletion, and the greenhouse effect. The depletion of nonrenewable and renewable resources (in the case of depletion beyond the point of reversibility) is an example of the impact of our domestic economic activities on future generations. Another example of the impact on future generations is the degradation of the environmental media: air, water, and soil. Current production and consumption levels and patterns, especially in the industrialized countries, cause serious environmental pollution, despite all the efforts to reduce emissions.

To be able to deal with these problems we have to overcome the market economic point of view. Every country has to accept responsibility for the impact of its economic activities on other countries, on the global ecosystem, and on future generations and must start to observe its economic activities from a responsibility point of view, considering the environment as an integral part of the economy.

AVOIDANCE COST APPROACH

The appropriate concept for such an attitude is the avoidance cost approach. This concept has been pioneered, developed, and put forward by such people as Daly, Hueting, and Ekins. It has also been elaborated in the U.N. *Handbook*. The additional cost item from the responsibility point of view are the costs of strategies necessary for achieving environmental sustainability.

The costs of the strategies or measures necessary to achieve environmental sustainability have to be calculated for the impact of domestic activities in the accounting period. Environmental sustainability means sustainable use of domestic and foreign natural capital, without affecting the global ecological system, maintaining natural capital intact to the extent that future generations will have the same opportunities to meet their needs as we have. Or, as Hueting says: using the environment (and resources) in such a way that the environmental functions remain available ad infinitum.

The strategies to avoid negative environmental impacts of economic activities differ with regard to the types of economic use of the natural environment:[15]

- With regard to the depletion of nonrenewable natural resources (such as subsoil assets), the quantitative decrease in these resources could be diminished by developing more efficient ways of using raw materials. Nevertheless, a decrease in these assets will generally be unavoidable. In this case, replacement by other

---

[15] This section has been drawn from Chapter 19, which has been provided by Carsten Stahmer.

types of natural capital might be necessary to achieve at least a constant level of natural capital as a whole (Daly, 1991). The substitution costs and/or the costs of efficiency improvements and the development, implementation, and operationalization of recycling systems for nonrenewables might be used as estimates of the environmental costs.

- With regard to the depletion of renewable cyclical natural assets, natural growth (biota), or natural inflow (groundwater) should be balanced against the quantities depleted. If depletion exceeds natural increase, the necessary reduction in net value added of the depleting industries can be used as an estimate of environmental costs.

- With regard to land use, sustainability implies a constant qualitative and quantitative level of landscapes and their ecosystems, including their biodiversity. If an increase in economic activities results in a decrease in this level, the necessary reduction in economic activities and in their net value added will have to be calculated.

- With regard to the discharge of wastes into nature, an analysis should be made of numerous possible prevention activities, comprising the development and implementation of end-of-the-pipe and process-integrated technology as a first step. If these activities are not sufficient to achieve the set sustainability standards, the next step is to decrease the level or restructure the pattern of economic activity. To reach specific sustainability standards, the strategy based on minimal costs should be chosen, with these prevention costs representing the deterioration of non-produced natural assets caused by pollution and so on.

Calculation of avoidance (prevention) costs is only possible on the basis of modelling. In the process of thinking about alternative and more sustainable modes of economic performance, any comparison between actual and desirable developments can only be hypothetical. The difference between the actual net domestic product and the hypothetical net domestic product could be interpreted as the necessary environmental costs. In this case, the eco domestic product would be the net domestic product of the hypothetical economy without negative impacts on the natural environment.

The methodological problems relating to the necessary modelling work have been elaborated in Chapters 14 and 15. Roughly speaking, two types of modelling can be distinguished: the static and the dynamic approach. In the static approach, the modelling exercise is based on current technology and current prices. In the dynamic approach one assumes changing (improving) technology and changing prices and preferences. In the static approach, the estimate of the costs to achieve environmental sustainability will probably be higher than in the dynamic approach, which implies a smoother transformation that is probably more realistic. The static approach makes the gap between the current level of activities and the level of activities in an environmentally sustainable economy more explicit. From a modelling point of view, the dynamic approach is much more complicated for it is simply not possible to predict economic, technological development and the changing preferences of economic subjects. Both approaches imply tricky assumptions.

Although there remain uncertainties and methodological problems, both types of economic modelling exercises can provide useful insights into the level and pattern of economic activities and the level of national income in a hypothetical, environmentally sustainable economy. Because of the many assumptions and the modelling characteristics, this approach cannot be implemented in the national accounts themselves. The adjusted domestic product, calculated by this approach, should at least be reflected in the SEEA. Even as a result of modeling; it is an important aggregate which should be linked to the traditional aggregate (GDP). It should be viewed as economic research on which statisticians and economists cooperate to gain insight into the consequences of achieving environmental sustainability at a given level and pattern of economic activity.

Recommendation 8

Although it is necessary to implement the market-based valuation approach, the comprehensive economic point of view is not sufficient to deal with all the questions relating to the achievement of environmental sustainability. The additional costs implied by adopting the responsibility point of view must also therefore be taken into account.

The appropriate concept for this attitude is the avoidance cost approach, which should be elaborated further to gain insight into the (costs of) strategies for achieving environmental sustainability. The avoidance cost approach requires modelling, which does not fit into the framework of the traditional SNA. Cooperation between statistical offices and economic research institutes is therefore necessary. Further research in this field should be carried out at the interface of such disciplines as national accounting, econometric modelling, and environmental economy. ∎

INTERNATIONAL TRADE

Until now we have concentrated on the depreciation of nonproduced natural assets in relation to domestic economic activities. The next extension is to introduce the environmental impact of international trade. The current balance of payments does not reflect any costs related to the environmental aspects of international trade. Doing so would imply a more comprehensive concept of responsibility.

In the case of international trade, responsibility for the environmental costs of traded goods and services should be borne by the country in which they are being consumed. For the domestic country, this means that the cost of changes in nonproduced natural assets caused by exported products should be deducted from the depreciation factor. Similarly, the costs of changes of nonproduced natural assets caused by imported products should be added to the depreciation factor.

Among other things, estimating the environmental costs of international trade could provide useful insights into the relationship between consumption in developed countries of products produced in developing countries.

Proposals to include the environmental aspects of international trade are also made in Chapter 15. Development of specific methodologies is at an early stage, at both the conceptual and the empirical level. Further research in this area is a matter of urgency, especially in view of the current trend towards globalization. Produced fixed capital is being located in those countries and regions of the world where wages are low and legal constraints limited. This encourages unsustainable development in the (developing) countries concerned, as we have explained at length in this book.

Recommendation 9

From a responsibility point of view, another extension of environmental cost accounting is to include the environmental aspects of international trade. This extension does not fit into the framework of the traditional SNA.

Environmental accounting that includes the environmental costs of international trade is at an early stage of development, at both the conceptual and the methodological level.

Considering the current trend towards globalization, there is an urgent need for further development of a framework and of methodologies for taking the environmental impact of international trade into account. Similar to the research on the avoidance cost method, this work should be carried out at the interface of such disciplines as national accounting and environmental economy. ■

8. RESISTANCE TO AND SOLUTIONS FOR CHANGE

The Club of Rome calls for the integration of environmental values in the System of National Accounts, instead of continuing the current practice which allows for consuming these resources and yet calls this activity income or growth. The reasons for change are ominous and need not be repeated here. At the same time, we are aware of the resistances to this change, which are assumed to be manifold. They are technical, ethical, and political in their very nature.

The notion that nature should be given a price often meets with an understandable resistance and even refusal. Whatever choices are made to correct GDP, they will always entail some calculation of the costs of depreciation, depletion, or simply of consumption of natural goods, in some way reflecting their estimated value. It is obvious that our society adheres to an ethic that holds that nature is God, and consequently that any pricing of nature is an act of blasphemy. This is an understandable reaction, to which we can only respond that advocates of this view must realize that they should then take a similar stand towards those who exploit God's nature, an act which is then, logically, also blasphemy.

Closely related to this kind of rejection is the conviction that the beauty of nature can never be valued at all, as this emotion is by definition subjective, and also that human values change with time and across cultures—so, any economic value is a construct. The transition to an economic system and theory which go beyond the traditional notion of economic (added) value requires acceptance of a certain degree of uncertainty as far as measurements are concerned (Giarini, 1993). This uncertainty stems from the fact that the very question of what wealth should be entails defining certain goals and expectations: the definition of a level of wealth is a function of time and history in evolution and, as such, a relative construct.

Another source of uncertainty in the notion of real wealth and welfare relates to the fact that many riches are conditioned by climatic conditions. Countries with cold climates will always need to develop more sophisticated heating systems than countries with milder climes. In the former, more monetarized activities have to be developed in order to provide artificial, manmade heat sources that can be stockpiled for winter. In milder areas, heating involves less provision and less expense. But which type of country is the poorer and which the richer: those which need to spend a lot of money on heating or those which need no heating at all?

We should never forget the paradox of heaven and hell: less scarcity leads quite naturally to less economic, monetarized wealth. However, where constraints are stronger, the stimulus to avoid hell in order to survive is probably greater. Many potentially poorer people have in the past become more industrious and richer than those who inhabited a more blessed environment. In all parts of the world, this is as true for individuals as it is for nations. But it is a historical process and it can be reversed. Furthermore, not all advantages are necessarily species-specific, for where life is exuberant and easy, it is so not only for the human species, but possibly also for competing biological beings like viruses.

Yet, as Max-Neef and Ekins have spelled out (1992), we can describe patterns of human behavior which are universal, and these include a sense of natural beauty, as well as the notion that clean air, water, soil, and so on are essentials of life.

Then, of course, there is formal resistance. In the first sessions of the U.N. Commission on Sustainable Development (1993), the US delegates made it clear that any changes in the international economic structure were off the agenda. The NAFTA treaty and the recently concluded round of GATT negotiations reflect the same pattern: that the world economic system is considered to represent a universal, and rigid, truth, any deviation from which gives rise to fear and menace. In the economic faculties of the world, little attention is paid to the environment or to the dynamics of the national income and the creation of welfare. In addition to this inertia there are the environmentalists, be they governments or NGOs, whom we generally know to cherish an allegiance with current economic growth patterns, largely because of their discomfort with the complexity of economic theories. And this is another point of resistance: the maze of theories in economics, not only in the field of GDP, but also in the principles underlying the economic behavior of states, companies, and consumers; discrepancies between the theory and practice of supply, demand, and trade, interpretations of income and costs, and the like. Yet, the theoretical framework for correcting national income can be considered sufficiently well developed to now make a start with formal adaptation.

Other resistances are to be expected that have their roots in the simplicity of the current GDP system, which seduces public opinion, so easily driven by the media and political response to messages of growth, being a one-dimensional account of production and consumption. As was observed by US leaders during recent elections, the average voter has only a superficial knowledge of the economy, as demonstrated by beliefs in possible relations between lower taxes and more defence, for example. The current trend of governmental retreat is another focus of concern. If the role of the state and international bodies is indeed to be diminished, then this is likely to slow down the momentum of any positive change in formal policies. And, finally, we have our doubts as to the willingness of governments and politicians to really strive for the truth, for honesty, for other than short-term profits and benefits, and to pick up the gauntlet of difficult long-term policies and measures requiring imagination and courage.

There is more to it, however, than "merely" correcting the GDP for its inadequacy to represent economic reality. Once the real index becomes accepted, what kind of new realities will then emerge in daily practice?

There is reason to believe that the myth of growth is so dominant that the logic of corrective measures as outlined in this book can and will not be followed unless we supply some additional answers. The answers on welfare or real wealth and on environmental capital have been given; those on the role of technology and employment not yet.

In every debate on this subject, we find ourselves confronted with the conviction that technology is there to make employment superfluous, this being the key to boost labor productivity. At the same time, market expansion is subsequently seen as the only way to compensate for the attendant redundancy of labor. In principle, there seems to be an endless spiral of new technologies which create (some) employment, which after a certain period has to be "rationalized", that is wiped out, after which another new technology and another round of labor redundancy follow. Market expansion, exports, and environmental exploitation supply the available instruments. If, for reasons of sustainability, this process is to be redirected, what will then happen to technology and employment?

Let us first realize that the aforementioned spiral is no longer creating sufficient employment, as neither the new markets nor the new technologies can keep pace with the redundancy process. Unemployment rates in Europe have mounted over the last 20 years from 2-3 percent (1970) to 7-8 percent (1980) and are at present 10-12 percent, all structural by definition, and expectations go beyond the scarcely imaginable figures of 20–25 percent. This means that, alongside adaptation of the GDP, something very fundamental has to be done to alter employment trends.

Our earliest message was to halt all growth, in the strict sense of the word. We do not withdraw that message, we rather expand it, by now calling for an efficiency revolution, which we believe will provide at least partial answers to the questions raised regarding technology and employment in relation to environmental constraints.

If we take the CO_2 issue as the key symbol of environmental impact, then we immediately crash into the technology issue. The Vostok expedition to the Antarctic has established a close correlation between CO_2 concentrations and global temperatures over the last 160,000 years (Figure 4). Even though the feedback loops are not yet well understood, we must assume that a further rise in concentrations of CO_2 (and other greenhouse gases) will raise global temperatures. The Intergovernmental Panel on Climate Change (IPCC) sees a reduction of greenhouse gas emissions by some 60 to 80 percent as essential for preventing a dangerous acceleration of global warming.

Contrasting sharply with the needs established by the IPCC, the

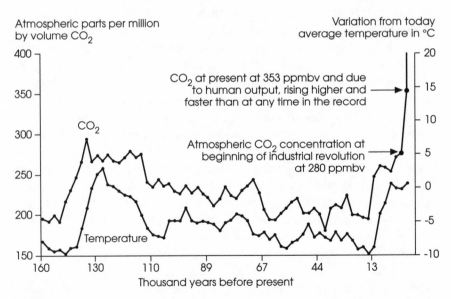

Figure 4   Correlation of CO_2 and temperature variation from 160,000 years before present to 2100

Source: Global Commons Institute, London.

scenarios of the World Energy Council (WEC) forecast an increase of energy demand worldwide by some 50 to 70 percent by 2020, which can be extrapolated to a doubling by 2040. Now we know the size of the gap that is opening between the exigencies of the IPCC and WEC scenarios: it is a factor 4 at least (Figure 5) Following the WEC scenarios, the resulting CO_2 emissions will be at least four times higher than the reduction of greenhouse gas emissions as proposed by the IPCC.

This gap cannot be closed by nuclear power. If we (very daringly) assume a tripling of nuclear capacity worldwide over the next 30 or 40 years, we would achieve an increase in the share of nuclear power in the world energy pie from 5, percent, today (Figure 6) to 15 percent (then). But if, in the meantime, the pie is doubling in size, that sector would not be worth more than 7.5 percent. This is simply too little to solve the problem. And all the unpleasant dangers accompanying nuclear power would be alarmingly increased, such as vulnerability to terrorism, uranium mining (which is unbelievably messy) and final waste disposal for hundreds of thousands of years.

Technological progress in the past has been characterized chiefly by the increase of labor productivity. Science and technology plus logistics and good management have enabled the most industrialized countries to increase labor productivity by a factor of roughly 20 over the past 150 years. In the process, resource productivity was left behind. It hardly increased at all, as is reflected in the fact that energy

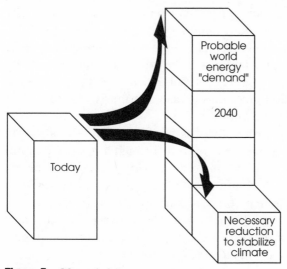

Figure 5 CO_2 emissions

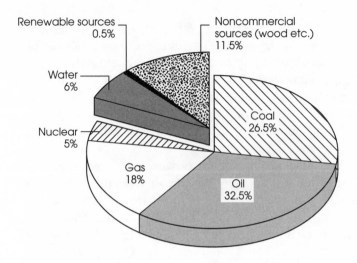

Figure 6   World primary energy consumption 1990

and material resource consumption grew almost parallel with GDP in all industrialized countries.

After 1973, a moderate delinking occurred, triggered by the increase in oil prices. But in the last 20 years this delinking has not boosted energy productivity by more than 20 percent in Japan, 10 percent in (West) Germany and probably not much more in the rest of Europe and the US.

The debate on the greenhouse effect seems to have shown that when it comes to macroeconomic energy productivity we should aim at more than a factor 4 (meaning an increase of 300 percent) to close the aforementioned gap. A factor 4 increase in energy productivity would allow a doubling of energy services while at the same time permitting energy-related greenhouse gas emissions to be halved. If fuel shifts are added, all the better.

Quadrupling energy productivity is not so outlandish as it may sound at first hearing. It will be achieved through a mere 3 percent annual increase over some 45 years. And for many processes involving energy consumption, a doubling of efficiency is possible even with existing technologies and without requiring any major changes in behavior or infrastructure. If these kinds of more comprehensive changes are allowed, a quadrupling of macroeconomic energy productivity is conceivable using existing technologies.

The point of entry for raising energy productivity will be Least

Cost Planning (LCP). If power plant permits are made dependent on the proof that there are no lower-cost alternatives available to fill the expected energy gap, it will soon become more profitable for utilities to subsidize energy efficiency at the consumer's end. If the monthly bills of their customers are reduced, utilities are permitted to raise the price of a kilowatt-hour. As a consequence, the capital costs will go down very substantially. Pacific Gas and Electric, the largest Californian utility, has scrapped its construction department entirely and is making more profits than before LCP was introduced.

The Wuppertal Institute is able to demonstrate the potential, in a wide variety of sectors, for quadrupling resource efficiency with no loss of service quality. We are concerned not only with energy but equally with material resource use, which at present is positively unsustainable (Schmidt-Bleek, 1994). In fact, the physical properties of materials as compared with energy make it even easier to conceive of huge rises in productivity. The key words here are longevity, remanufacturing, recycling and reduced transportation needs. What are clearly lacking are rational policies to reduce material flows. We have had no equivalent to the energy crisis and hence lack 20 years of valuable experience.

The easiest way of driving technology development in the new direction is to let the prices speak the "ecological truth" (or nearly) and thereby render the desired technological revolution profitable at the company level as well as for final consumers.

The macroeconomic justification for doing so comes from estimates of negative externalities caused by resource use. Prognos, the Basel-based consultancy firm, has estimated the external costs of energy production and use in Germany at some 5–10 percent of the German GDP (Prognos, 1992). This extensive study was paid for by the German ministry of economic affairs, and it points to the magnitude of corrected GDP figures.

Price increases for energy and primary raw materials can be set in motion by a variety of measures, including technical and bureaucratic stipulations, pollution control laws, the cutting of subsidies, tradable permits (e.g., for CO_2 emissions) and, finally, taxes and charges.

In the international arena, tradeable CO_2 permits (e.g., Grubb, 1992), issued on a per-adult basis, would be our favorite tool. They would clearly benefit the South but still exert strong pressure on that hemisphere to keep its energy consumption low. Some concessional "grandfathering" (i.e. special allowances for the traditional big consumers) could be negotiated and might be justified as a way of recognizing the difficulties of sudden adaptation.

Tradeable permits are less plausible for domestic environmental policies. Who is there to measure and control CO_2 emissions from wood stoves or methane emissions from compost? Moreover, permit prices are likely to increase sharply in times of economic boom, causing hardship to millions of people who may not enjoy any of the benefits of the economic upswing. By comparison, all the problems of administrative costs, unpredictability of prices and social equity can be greatly reduced by a strategy of direct pricing of CO_2 emissions or of energy, which we are advocating under the name of ecological tax reform.

Ecological tax reform is not an unequivocal concept. There are certainly many ways of creating a lot of damage through an ill-designed tax reform or through ill-bred implementation.

A few hints may help in finding an appropriate design.

1. No quick ecological effects should be expected from green taxation. The short-term price elasticities of energy and other resources are very low. This fact should not be misconstrued as pointing to the poor efficiency of the instrument. Long-term price elasticities can be expected to be very high, even for car fuels, the example often quoted for nearly non-existent price elasticities. A comparison among OECD countries shows that there is a strong negative correlation between fuel prices and annual per capita fuel consumption.

2. Because short-term price elasticities are low and long-term price elasticities are high, there is no point in introducing green taxes very quickly. To achieve the desired steering effect it is far more important to create an indisputable certainty of a gradual rise in green taxes over decades. To be effective, it is unnecessary to make a prompt start with $3 per barrel. The initial price signal could be much smaller or even zero if there is a certainty that the taxation will be introduced and will reach levels of $10 per barrel at least, and ultimately much more. It is a perfectly legitimate demand on the part of industry that the signal should not lead to capital destruction (premature closure of existing facilities). But it is not legitimate to ask for conditions encouraging fresh money to flow into unsustainable structures.

3. Revenue neutrality is essential for positive employment effects. State spending tends to be less effective in job creation than private spending under favorable conditions. For the greatest impact on job creation, indirect labor costs should be reduced with the highest priority. A reduction in social security payments, health insurance premiums, and similar indirect labor costs would make human labor more affordable for employers. Obviously, the proceeds of green taxation would then have to be used in part to co-finance the respective services.

4. There are no ecological arguments against revenue neutrality. Earmarking of the proceeds is unnecessary. The strongest ecological effect of ecological tax reform comes from the change in the framework of profitability. If there are strong additional reasons to spend more public money on environmental protection or on mass-transport infrastructure, parliaments are free to decide so—but at the expense of other sectors.

Reflecting and considering these observations, we advocate Von Weizsäcker's proposal for ecological tax reform, which we consider acceptable from the point of view of a society which has adopted employment as its first priority, namely a slow raising of resource prices by some 5 percent annually over a long period of perhaps some 40 years (Von Weizsäcker, 1994). This can be achieved first by cutting subsidies on energy (and, likewise, on other ecologically problematic factors). Subsequently, taxes could be levied on nonrenewable sources of energy, on primary raw materials, on water consumption, on certain chemicals such as chlorine or metals, and on certain types of land use. We remain skeptical about taxes on polluting emissions or on waste, owing to problems of evasion and control, chiefly in less developed countries.

Other taxes, charges, and levies should be reduced by equivalent amounts. And in particular the fiscal burdens on human labor should be reduced. Especially in the European fiscal system, taxation on labor is such that incomes suffer, and making labor redundant seems to have become a major incentive for employers. The plea for ecotaxes (or an energy tax) in Europe is therefore counterbalanced by a relief in labor tax.

The cutting of subsidies plus ecological tax reform will make the increase of resource productivity more profitable and the dismissal of workers less. This should be good for the economy as a whole. Repetto et al. (1992) believe that the total possible gain from shifting to environmental charges could easily be $0.45 to $0.80 per dollar of tax shifted from "goods" to "bads"—with no loss of revenues.

Owing in part to the expected productivity gains (estimated conservatively at 3 percent annually), the price signal of 5 percent annually would be extremely tame. It would effectively be 2 percent (5 minus 3 percent) annually for a production factor accounting on average for less than 4 percent of overall production costs, so that the ultimate cost differential would be only 0.08 percent annually. Even this almost imperceptible cost differential would—on average—be outbalanced by labor cost reductions, e.g. from reduced social security payments. Hence firms would on average even benefit financially from the reform. In other words, even if some businesses may suffer losses, on average, there would be more winners than losers.

Obviously, the main problems are predictability and international harmonization. But if there is a societal consensus that the scheme is beneficial for the economy, and not only for the environment, it should not be too difficult to reach agreement on it among all major political parties and thus keep it out of election campaigns after a given time. International harmonization will be much easier than in the case of classical environmental policy measures, because the latter have invariably involved extra costs with no immediate economic benefits.

If business leaders and capital owners see scope for a new and reliable avenue of technological progress, they may feel encouraged to pour money into the new trend. There is virtually nothing more stimulating for the business world than a steady and predictable trend. Once the consensus is strong enough to persuade the pioneers, the process can soon become self-enforcing. Durable goods would become obsolescent somewhat earlier. This could easily serve as a powerful initial kickstart, strengthening the still vulnerable economic upswing.

Is it not an extremely tempting idea to use the high certainty of the environmental crisis as the basis for a high certainty of a new technological trend? Is it not, therefore, a good idea to put the shift of technological progress on the agendas of the G7, of OECD, or of the relevant U.N. bodies?

The world would look quite different from today's. Politicians would no longer fear the shrinking of the pie of formal employment. "Growth" would no longer be a prerequisite for political stability or for stable public budgets. Indicators of real wealth would no longer be an academic concept, but something to be established empirically by close observation of the marketplace. This is because the job market would then be largely a market for trading opportunities for real satisfaction.

One could go even further in designing the sustainable society. Much of what was called ultimate satisfaction can be obtained with negligible amounts of the commodities mentioned. Commercial energy may be completely dispensable for some communities insisting on self-sufficiency. Education systems in the sustainable society would be fundamentally different from today's. The same is true of the traffic system, the police, and many other public functions.

We have walked this considerable distance towards utopia deliberately. We want to express our opinion that achieving sustainability is not only a matter of measurement and academic definition. It is essentially a matter of forcefully redirecting technological and cultural progress, starting with appreciable modifications of our economic incentive systems. Ultimately we shall need a vision of a new set of values and a new texture to civil society.

9. STRENGTHENING INTERNATIONAL COORDINATION[16]

In the last ten years, the international community has joined forces to estimate environmental values within the framework of national accounts. Several international agencies, noticeably the UNEP, U.N. Statistical Division (UNSTAT), and the World Bank, together with a few governments, have sponsored workshops that have prompted research activities. Academic institutions and NGOs have also participated in some of these activities. The results of these efforts are reflected in various publications on conceptual issues, methodologies, and case studies, including the U.N. *Handbook on a System for Integrated Environmental and Economic Accounting (SEEA),* which provides a framework for data classification, accounting standards and environmental valuation methodologies. These efforts, however, fall

---

[16] For more information on the history of failures of the SNA, national and international efforts on environmental accounting, and the WWFF view on the SNA reform, see "Real Value for Nature," a WWF-International publication (Sheng, 1995).

far short of what is required for integrating environmental values into the System of National Accounts (SNA).

To bring the environment into the SNA and into decision-making processes requires more than a few workshops and publications. It requires public support and political will, an internationally comparable framework, financial and technical assistance to developing countries for implementing the integration, and most of all, a strong international leadership that sets the direction, guides the course, mobilizes funds and expertise, supports national efforts, organizes and coordinates research activities, disseminates information, and applies a reformed SNA at the international level. Coordinated international efforts on these fronts, which are presently lacking, are the measure of the seriousness of the exercise. Despite the slowly growing number of publications on the subject, there is a lack of comprehensible and accessible information for the public, whose awareness and opinions are among the basic variables determining the political will of policymakers. Most publications are highly technical, with no explanation of the basic rationale for reforming the SNA. There is no targeted distribution of the publications. Even some of the most relevant government agencies, such as environment departments or ministries of economic planning, particularly those in the developing world, do not have their names on the distribution list, not to mention the nongovernmental sectors. This situation is slowing the momentum of the reform. Despite the existence of several methodologies for environmental accounting, there is a lack of open debate on them, thus hindering their improvement and harmonization and stalling the development of an internationally comparable framework which gives meaning to the exercise. Individual experts tend to be protective of the methodologies they have developed and allow little leeway for modifications or for accepting the merits of the methodologies developed by others. There is little international effort to assess the methodologies. In Indonesia, six methodologies have been tried without conclusions being drawn as to their pros and cons. The continued divergence of the methodologies not only leads to inefficient use of the scarce research resources, but also slows down the reform process in that it encourages policymakers to take a "wait-and-see" attitude. Policymakers have a point: the adoption of any methodology is an expensive exercise. International agencies have provided financial and technical support to a limited number of countries, but they have made no major commitments to international application of an environmentally adjusted SNA.

U.N. Agencies face budgetary constraints across the board. The U.N. Economic Commission for Europe (ECE), one of the few agencies specializing in environmental statistics, is suffering severe budget cuts. Multilateral development banks (MDB) have the strongest financial and technical capacity, yet their support is not forthcoming. The Organization of American States (OAS) has developed a Western Hemisphere Program on environmental accounting which requires US$3 million over a period of three years for networking and capacity building in its developing member countries. It has approached a few MDBs and received no positive response. Most importantly, the international leadership required for the reform of an international accounting system such as the SNA is lacking. The loosely organized Inter-Secretariat Working Group (ISWG) on the revision of the SNA has no clear mandate on the issue of environmental accounting, although some environmental elements have been brought to the periphery of the system in the 1993 version of the SNA. The ISWG is incapable of setting an international agenda and coordinating the reform process. The World Bank used to play a key role in the 80s, in terms of organizing research activities and providing financial and technical input, but the lack of political will on the part of the Bank's major shareholders has prevented the Bank from even keeping the subject on its research agenda. U.N. Agencies are constrained by their institutional capacity and political influence, which are necessary for enforcing the application of an environmentally adjusted SNA at the global level. The U.N. *Handbook,* which represents a satellite approach leaving the existing SNA intact, is partly a reflection of such constraints. Lack of political will on the part of the member governments have prevented OECD and the European Union from giving priority to the subject, though important initiatives from the side of the European Commission and Parliament have been set on track recently (see also the end of Chapter 9). In summary, the absence of coordinated international efforts in terms of information dissemination, public debate, financial and technical assistance, and international leadership is weakening public support and political will, impeding the development of an internationally comparable framework, failing to build up national capacities for implementation, and misdirecting the process of the reform. To be serious about integrating environmental values into the SNA, the international community must take immediate actions in the following areas.

1. Disseminating related information in a comprehensible form to governments, development agencies, national statistical offices,

research institutes, the NGO community, and other concerned parties. Providing education to the public and policymakers on the necessity and possibility of reforming the existing SNA.

2. Facilitating open debates among experts and interested parties on methodological issues and on policy application of an environmentally adjusted SNA.

3. Moving beyond the satellite approach of the U.N. *Handbook,* integrating environmental accounting into the SNA core accounts, and developing an internationally comparable framework for measuring the economic aspects of sustainability, starting by adjusting national income for the most serious distortions.

4. Strengthening coordination among international agencies in research activities, pilot projects, national or regional focus, and application of adjusted economic indicators in the publications, official documents, macroeconomic analysis, and design of the development programs and projects of each agency.

5. Linking environmental data collection with monetary valuation to the extent possible and facilitating dialogues among natural scientists, national accountants, economists, and statisticians, to enhance their collaboration.

6. Providing financial, technical, and institutional support to national statistical offices, in such areas as data collection, compilation of environmental physical accounts (particularly for the most seriously threatened resources) natural resource valuation, environmental monitoring, training of national accountants, pilot projects, and the creation of regional and subregional centers of expertise.

7. Including environmental accounting in national sustainability strategies and reports and using adjusted economic indicators in decision-making processes and for guiding economic policies. Applying environmental accounting techniques at micro and international economic levels, i.e., in cost/benefit analysis of projects and in international balance of payments calculations.

8. Reorganizing the ISWG to expand it into an international working group involving not only major international agencies, but also national statistical offices, research institutes, NGOs, and other interested parties. It should be responsible for developing

a global agenda, coordinating and monitoring international efforts, and enforcing the implementation of an environmentally adjusted SNA. It should report to the U.N. Commission on Sustainable Development (CSD), which should in turn publicize the work of this group on a regular basis.

10. SUMMARY OF CONCLUSIONS AND RECOMMENDATIONS

- It is necessary to develop more comprehensive measures of societal progress and to improve GDP and NDP as indicators for economic progress.
- A wide variety of methods have been proposed to allow society to make a more comprehensive assessment of actual societal developments. As yet, most of these proposals have been developed independently of one another. It would be worth the effort to establish an international working group to make an inventory and critical assessment of the available approaches, making explicit their similarities and differences, and respective strengths and weaknesses. The next step could be to develop a more generally accepted framework, for both the basket approach (set of indicators) and the index approach. Despite the present absence of such a framework, data are available on most of the welfare components mentioned above. We therefore call on national statistical offices, politicians, and the media not to publish GDP in isolation from, but in the context of other economic, social, and environmental indicators.
- Agenda 21 stresses the need for developing a consistent framework of economic, social, and environmental indicators for Sustainable Development. These indicators will enable governments and supranational organizations to assess the extent to which societies are on a sustainable path. The establishment of standards should be based on intra- and intergenerational equity.
- The U.N. *Handbook on Integrated Environmental and Economic Accounting* should be implemented by the national statistical offices as soon as possible. This process should be stimulated more intensively by UNSTAT, World Bank, and other

international agencies involved in the SNA (European Union, OECD, IMF).

- From a comprehensive economic point of view, net (environmentally adjusted) value added = gross output at market values minus intermediate inputs minus the depreciation of produced fixed assets minus the depreciation[16] of nonproduced natural assets (the costs of economic input in the domestic country during the accounting period). The depreciation of nonproduced natural assets must be excluded from GDP. This can be done by deducting the value changes as a depreciation factor or by including the consumption of these assets as intermediate inputs. To reflect the depreciation of nonproduced natural assets at market values in the flow accounts is a logical and necessary first step in improving the SNA and should be implemented as soon as possible. The new NDP could be calculated as follows:

    Old GDP ÷ Costs of depleting inventories (user cost approach)
    = New GDP
    ÷ Costs of depreciation of manmade capital (net present value method)
    ÷ Costs of depreciation of natural assets (net price method)
    ÷ Costs of degradation (replacement cost method)
    = New NDP (EDP = NDP in transition)

    Among the experts, there is no agreement on the proper market valuation methodology. We call on the experts in this area to reach consensus on the proper market valuation methodology, which can then be endorsed by the relevant international agencies involved in the revision of the SNA (UNSTAT, World Bank, IMF, EU, OECD). The statistical offices can then experiment and gain experience by first introducing this recommendation in satellite accounts. In the next SNA revision, these value changes of nonproduced natural assets should be reflected in the Core System as well.
- From a comprehensive economic point of view, physical changes in natural assets during the accounting period within

---

[17] Depreciation: depletion and degradation.

the domestic country which have no market value are not taken into account. Nevertheless, as these asset changes represent an additional cost item, they should be excluded from value added.

- Given the state of the art at the conceptual and methodological level, the choice made in the U.N. *Handbook,* namely to show the environmental defensive expenditures as a ratio, rather than to deduct them, can be defended. However, before making a final decision on how to deal with these expenditures in relation to GDP/NDP, the issues mentioned here need clarification by arranging expert meetings on them.
- Although it is necessary to implement the market-based valuation approach, the comprehensive economic point of view is not sufficient to deal with all the questions relating to the achievement of environmental sustainability. The additional costs implied by adopting the responsibility point of view must also therefore be taken into account.

The appropriate concept for this attitude is the avoidance cost approach, which should be elaborated further to gain insight into the (costs of) strategies for achieving environmental sustainability. The avoidance cost approach requires modelling, which does not fit into the framework of the traditional SNA. Cooperation between statistical offices and economic research institutes is therefore necessary. Further research in this field should be carried out at the interface of such disciplines as national accounting, econometric modelling, and environmental economy.

- From a responsibility point of view, another extension of environmental cost accounting is to include the environmental aspects of international trade. This extension does not fit into the framework of the traditional SNA. Environmental accounting that includes the environmental costs of international trade is at an early stage of development, at both the conceptual and the methodological level.

Considering the current trend towards globalization, there is an urgent need for further development of a framework and of methodologies for taking the environmental impact of international trade into account. Similar to the research on the avoidance cost method, this work should be carried out at the interface of such disciplines as national accounting and environmental economy.

References

Chapter 1

Daly, H., and J. Cobb (1989), *For the Common Good*, Beacon Press, Boston.

Daly, H., and J. Cobb (1994), *For the Common Good*, Second edition, Beacon Press, Boston.

Diefenbacher, H. (1991), "Der 'Index of Sustainable Economic Welfare'. Eine Fallstudie über die Entwicklung in der Bundesrepublik Deutschland," in Diefenbacher, H., und S. Habicht-Erenler (eds.), *Wachstum und Wohlstand*, Metropolis Verlag, Marburg.

Diefenbacher, H. (1994), "The Index of Sustainable Economic Welfare—A Case Study of the Federal Republic of Germany," in Daly, H., and J. Cobb (1994), *For the Common Good*, second edition, Beacon Press, Boston.

Giarini, O., and W.R. Stahel (1993), *The Limits to Certainty*, Kluwer Academic Publishers, Dordrecht.

Jackson, T., and N. Marks (1994), *Measuring Sustainable Economic Welfare (UK)—A Pilot Index: 1950-1990*, Stockholm Environmental Institute, Stockholm.

Kaplan, R.D. (1994), *Het anarchistisch pandemonium*, NRC Handelsblad, Zaterdags Bijvoegsel (14 mei) (© The Atlantic Monthly, Boston MA).

Kennedy, P. (1993), *Preparing for the Twenty-First Century*, Fontana Press, London.

Meadows, D.H., D.L. Meadows, and J. Randers (1991), *Beyond the Limits: Confronting Global Collapse: Envisioning a Sustainable Future*, Earthscan, London.

Obermayr, B., K. Steiner, E. Stockhammer, und H. Hochreiber (forthcoming in 1994), *Die Entwicklung des ISEW in Österreich von 1955 bis 1992*.

Rosenberg, D., and T. Oegema (forthcoming in 1994), *A Pilot Index of Sustainable Economic Welfare for the Netherlands, 1950–1992,* IMSA, Amsterdam.

CHAPTER 2

Delumeau, J. (1978), *La peur en Occident,* Librairie A. Fayard, Paris.

Hobbes, Th. (1983), *Leviathan,* Everyman's Library, London (1651).

Huizinga, J. (1963), *Herfsttij der Middeleeuwen,* Tjeenk Willink en Zn., Haarlem.

Keynes, J.M. (1930), *Essays in Persuasion,* W.W. Norton, New York and London.

Locke, J. (1960), *Two Treatises of Government,* Cambridge University, Cambridge (1690).

Mill, J.S. (1985), *Principles of Political Economy,* Penguin Books Harmondsworth (1848).

Toulmin, S. (1990), *Cosmopolis: the Hidden Agenda of Modernity,* University of Chicago Press, Chicago.

Xenos, N. (1989), *Scarcity and Modernity,* Routledge, London and New York.

CHAPTER 3

Algera, S.B., and R.J.A. Janssen (1991), "The development of short-term macro-economic statistics," *CBS Select,* 7.

Bochove, C.A. van, and H. K. Van Tuinen (1985), "Revision of the System of National Accounts: the case for flexibility," *National Accounts Occasional Paper,* NA/06.

Eck, R. van, C. N. Gorter, and H. K. Van Tuinen (1983), "Flexibility in the System of National Accounts," *National Accounts Occasional Paper,* NA/01.

Haan, M. de, S. Keuning, and P. Bosch (1993), "Integrated indicators in a National Accounting Matrix including environmental accounts (NAMEA); an application to the Netherlands," *National Accounts Occasional Paper,* NA/60).

King, G. (1936), *Two Tracts,* The John Hopkins Press, Baltimore.

Metelerkamp, R. (1804), *De toestand van Nederland in vergelijking gebracht met die van enige andere landen van Europa,* Rotterdam.

Robinson, A. (1986), "A child of the times," *The Economist* (20 December).

CHAPTER 4

Bones, J. (1993), branch chief, Forest Inventory Research and Analysis, Forest Service, USDA, Washington D.C., personal communication, October 22.

Canadian Council of Forest Ministers (1993), *Compendium of Canadian Forestry Statistics 1992,* National Forestry Database, Ottawa.

Daly, H.E. (1992): "Allocation, Distribution, and Scale: Towards an Economics That is Efficient, Just, and Sustainable," *Ecological Economics,* (December).

Dregne, H. et al. (1991), "A New Assessment of the World Status of Desertification," *Desertification Control Bulletin,* No. 20.

FAO (United Nations Food and Agriculture Organization) (various years), *Yearbook of Fishery Statistics: Catches and Landings,* Rome.

FAO (United Nations Food and Agriculture Organization) (1988a), "Areas of Woody Vegetation at End 1980 for Developing and Developed Countries and Territories by Region," in *An Interim Report on The State of Forest Resources in the Developing Countries,* Rome.

FAO (United Nations Food and Agriculture Organization) (1988b), *An Interim Report on the State of Forest Resources in the Developing Countries,* Forest Resources Division, Rome.

FAO (United Nations Food and Agriculture Organization) (1991), *Food Balance Sheets* , Rome.

FAO (United Nations Food and Agriculture Organization) (1992), *Production Yearbook 1991,* Rome.

FAO (United Nations Food and Agriculture Organization) (1992 and 1993), *Forest Resources Assessment 1990,* Rome.

FAO (United Nations Food and Agriculture Organization) (1993a), *AGROSTAT-PC 1993*, Forest Products Electronic Data Series, Rome.

FAO (United Nations Food and Agriculture Organization) (1993b), "Crops from Pasture Land," *International Agricultural Development* (March/April).

FAO (United Nations Food and Agriculture Organization) (1993c), *Forest Resources Assessment 1990 — Tropical Countries,* Forestry Paper No. 112, Rome.

FAO (United Nations Food and Agriculture Organization) (March 1993d), *Global Fish and Shellfish Production in 1991*, Fisheries Department, COFI, Support Document: Fishery Statistics, Rome.

FAO (United Nations Food and Agriculture Organization) (1993e), "Marine Fisheries and the Law of the Sea: A Decade of Change," *Fisheries Circular*, No. 853, Rome.

FAO (United Nations Food and Agriculture Organization) (1993f), "World Review of High Seas and Highly Migratory Fish Species and Straddling Stocks," *FAO Fisheries Circular*, No. 858 (preliminary version), Rome.

FAO (United Nations Food and Agriculture Organization) (1993g), *1961-1991 ... 2010: Forestry Statistics Today for Tomorrow,* Rome.

Gulland, J.A. (ed.) (1971), *The Fish Resources of the Ocean*, Fishing News (Books) Ltd., West Byfleet, Surrey, U.K.

He Bochuan (1991), *China on the Edge: The Crisis of Ecology and Development,* China Books and Periodicals, Inc., San Francisco.

Houghton, R.A., et al. (1983): "Changes in the Carbon Content of Terrestrial Biota and Soils Between 1860 and 1980: A Net Release of CO_2 to the Atmosphere," *Ecological Monographs* (September).

Janz, K. (1993), senior forestry officer, Resources Appraisal and Monitoring, Forestry Department, FAO, Rome, private communication, October 29.

Lanly, J.P. (1983), *1980 Forest Resources Assessment,* FAO, Rome.

Lowe, J. (1993), Forest Inventory and Analysis Project, Petawawa National Forestry Institute, Canadian Forest Service, Chalk River, Ontario, unpublished printout and personal communication, November 4.

Malik, R.P.S., and P. Faeth (1993): "Rice-Wheat Production in Northwest India," in Faeth, Paul (ed.), *Agricultural Policy and Sustainability: Case Studies from India, Chile, the Philippines, and the United States,* World Resources Institute, Washington, D.C..

Meadows, D.H., D.L. Meadows, and J. Randers (1991): *Beyond the Limits. Confronting Global Collapse; Envisioning a Sustainable Future,* Earthscan Publications Ltd., London.

National Commission on the Environment (July 1991), National Report to the United Nations Conference on Environment and Development, Republic of Argentina.

Oldeman, L.R. et al. (1991), "The Extent of Human-Induced Soil Degradation," in Oldeman, L.R. et al., *World Map of the Status of Human-Induced Soil Degradation,* United Nations Environment Program and International Soil Reference and Information Center, Wageningen, the Netherlands.

Perotti, M. (1993), chief, Statistics Branch, Fisheries Department, FAO, Rome, personal communication, November 3.

Postel, S., and J.C. Ryan (1991), "Reforming Forestry,"in Brown, L.R. et al., *State of the World 1991,* W.W. Norton & Co., New York.

Postel, S. (1992), *Last Oasis: Facing Water Scarcity,* W.W. Norton & Company, New York.

Shvidenko, A. (1993), International Institute for Applied Systems Analysis (IIASA), Laxenburg, Austria, unpublished printout and personal communication, July 18.

Smil,V. (1992), "China's Environment in the 1980s: Some Critical Changes," *Ambio,* September.

Strand, J. and F. Wenstöp (. . .) *Kvantifisering av Miljölemper ved udlike energiteknologier,* Sosialökonomisk Institut, Oslo University, Oslo.

Thomas, R.J. (1993), Centro Internacional de Agricultura Tropical, Cali, Colombia, personal communication, July 21.

Tuijl, W. van (1993), *Improving Water Use in Agriculture: Experiences in the Middle East and North Africa,* World Bank, Washington, D.C.

UN Economic Commission for Europe/FAO (1992), *The Forest Resources of the Temperate Zones: the UN-ECE/FAO 1990 Forest*

Resource Assessment, Vol. 1: General Forest Resource Information, United Nations, New York.

United States Bureau of the Census (1993), *International Data Base*, Department of Commerce, November 2 (unpublished printout).

United States Department of Agriculture (1982), *An Analysis of the Timber Situation in the United States: 1952-2030*, Forest Service, Forest Resource Report No. 23, Washington, D.C.

United States Department of Agriculture (1992), *World Grain Database*, Washington, D.C. (unpublished printout).

Vitousek, P.M. et al. (1986), "Human Appropriation of the Products of Photosynthesis," *BioScience* (June).

Waddell, K., D. Oswald, and D. Powell (1989), Pacific Northwest Experiment Station, Forest Service, *Forest Statistics of the United States 1987*, Research Bulletin 168, Portland, Oreg. USDA.

Wardle, P. (1993a), Senior Forestry Economist, FAO, Rome, Italy, unpublished printout and personal communication, July 1.

Wardle, P. (1993b), Senior Forestry Economist, FAO, Rome, Italy, unpublished printout from FAO Forestry Products Database, September 21

Chapter 5

Adger, N. (1992), *Sustainable National Income and Natural Resource Degradation: Initial Results for Zimbabwe*, CSERGE GEC Working Paper 92-32, University of East Anglia, Norwich.

Alfsen, K., T. Bye, and L. Lorentsen (1987), *Natural Resource Accounting and Analysis: the Norwegian Experience 1978-1986*, Central Bureau of Statistics of Norway, Oslo.

Bartelmus, P., E. Lutz, and S. Schweinfest (1993), "Integrated Environmental and Economic Accounting: a Case Study for Papua New Guinea," in Lutz, E. (ed.) (1993), *Toward Improved Accounting for the Environment*, World Bank, Washington D.C., pp. 108-143.

Bartelmus, P., C. Stahmer, and J. Van Tongeren (1993), "Integrated Environmental and Economic Accounting—a Framework for an SNA Satellite System," in Lutz, E. (ed.) (1993), *Toward Improved*

Accounting for the Environment, World Bank, Washington DC, pp. 22-45.

Bartelmus, P., and A. Tardos (1993), "Integrated Environmental and Economic Accounting—Methods and Applications," *Journal of Official Statistics* Vol. 9, No.1, pp. 179-188.

Beckerman, W. (1968), *An Introduction to National Income Analysis*, Weidenfeld & Nicholson, London.

Boero, G., R. Clarke, and L. Winters (1991), *The Macroeconomic Consequences of Controlling Greenhouse Gases: a Survey*, Department of the Environment, HMSO, London.

Bojö, J., K.-G. Mäler, and L. Unemo (1992), *Environment and Development: an Economic Approach*, Kluwer, Dordrecht.

Bryant, C., and P. Cook (1992), "Environmental Issues and the National Accounts," *Economic Trends* No.469, November, HMSO, London, pp. 99-122.

Cobb, C.W., and J.B. Cobb (1994), *The Green National Product*, University Press of America, Lanham MD.

Daly, H.E., and J.B. Cobb (1990): *For the Common Good*, Green Print, Merlin Press, London.

Eisner, R. (1988), "Extended Accounts for National Income and Product," *Journal of Economic Literature* Vol.26 (December), pp. 1611-1684.

Ekins, P. (forthcoming in 1994), "The Environmental Sustainability of Economic Processes," in Van den Bergh, J., and J. van der Straaten: *Concepts, Methods and Policy for Sustainable Development*, Island Press, Washington D.C.

Ekins, P., and M. Max-Neef (eds.) (1992), *Real-Life Economics: Understanding Wealth Creation*, Routledge, London.

El Serafy, S. (1989), "The Proper Calculation of Income from Depletable Natural Resources," in Ahmad, Y., S. El Serafy, and E. Lutz (eds.) (1989), *Environmental Accounting for Sustainable Development*, World Bank, Washington D.C., pp. 10-18.

El Serafy, S. (1993), "The Environment as Capital," in Lutz, E. (ed.) 1993: *Toward Improved Accounting for the Environment*, World Bank, Washington D.C., pp. 17-21.

European Commission (1992), *Towards Sustainability: a European Community program of policy and actions relating to the environment and sustainable development*, Vol.II, EC, Brussels.

Goldschmidt-Clermont, L. (1992), "Measuring Households' Non-Monetary Production" in Ekins, P., and M. Max-Neef, (eds.) (1992), *Real-Life Economics: Understanding Wealth Creation*, Routledge, London, pp. 265-280.

Goldschmidt-Clermont, L. (1993), "Monetary Valuation of Non-Market Productive Time: Methodological Considerations," *Review of Income and Wealth* Series 39 No.4 (December), pp. 419-433.

Harrison, A. (1993), "National Assets and National Accounting," in Lutz, E. (ed.) (1993), *Toward Improved Accounting for the Environment*, World Bank, Washington D.C., pp. 22-45.

Herfindahl, O., and A. Kneese (1973), "Measuring Social and Economic Change: Benefits and Costs of Environmental Pollution," in Moss, M. (ed.) (1973), *The Measurement of Economic and Social Performance*, National Bureau of Economic Research/Columbia University, New York, pp. 441-508.

Hicks, J. (1946), *Value and Capital*, Second edition, Oxford University Press, Oxford.

Hueting, R. (1986), "An Economic Scenario for a Conserver Economy," in Ekins, P. (ed.) (1986), *The Living Economy: a New Economics in the Making*, Routledge & Kegan Paul, London, pp. 242-256.

Hueting, R. (1989), "Correcting National Income for Environmental Losses: Toward a Practical Solution" in Ahmad, Y., S. El Serafy, and E. Lutz (eds.) (1989), *Environmental Accounting for Sustainable Development*, World Bank, Washington D.C.

Hueting, R., P. Bosch, and B. de Boer (1991), *Methodology for the Calculation of Sustainable National Income*, Netherlands Central Bureau of Statistics, Voorburg.

Jorgenson, D. (1990), "Productivity and Economic Growth," in Berndt, E.R., and J.E. Triplett (eds.) (1990), *Fifty Years of Measurement: the Jubilee of the Conference on Research in Income and Wealth*, University of Chicago Press Chicago, pp. 19-118.

Jorgenson, D., and Wilcoxen (1993), "Energy, the Environment and Economic Growth," in Kneese, A., and J.L. Sweeney (eds.)

(1993),*Handbook of Natural Resource and Energy Economics*, Volume 3, North-Holland, Amsterdam, pp. 1267-1349.

Juster, F.T. (1973), "A Framework for the Measurement of Economic and Social Performance," in Moss, M. (ed.) (1973), *The Measurement of Economic and Social Performance*, National Bureau of Economic Research/Columbia University, New York, pp. 25-109.

Kuznets, S. (1973), "Concluding Remarks," in Moss, M. (ed.) (1973), *The Measurement of Economic and Social Performance*, National Bureau of Economic Research/Columbia University, New York, pp. 579-592.

Leipert, C. (1989a), "National Income and Economic Growth: the Conceptual Side of Defensive Expenditures," *Journal of Economic Issues* Vol. XXIII No. 3, (September).

Leipert, C. (1989b), "Social Costs of the Economic Process and National Accounts: the Example of Defensive Expenditures," *Journal of Interdisciplinary Economics* Vol. 3, pp. 27-46.

Lone, Ø. (1992), "Environmental and Resource Accounting" in Ekins, P., and M. Max-Neef (eds.) (1992), *Real-Life Economics: Understanding Wealth Creation*, Routledge, London, pp. 239-253.

Lutz, E. (1993), "Toward Improved Accounting for the Environment: an Overview," in Lutz, E. (ed.) (1993), *Toward Improved Accounting for the Environment*, World Bank, Washington D.C., pp. 22-45.

Mäler, K.-G. (1991), "National Accounts and Environmental Resources," *Environmental and Resource Economics* Vol. 1, pp. 1-15.

Miles, I. (1992), "Social Indicators for Real-Life Economics," in Ekins, P., and M. Max-Neef (eds.) (1992), *Real-Life Economics: Understanding Wealth Creation*, Routledge, London, pp. 283-297.

Nordhaus, W. (1991), "To slow or not to slow: the economics of the greenhouse effect," *Economic Journal* 101 (July), pp. 920-937.

Nordhaus, W., and J. Tobin (1973), "Is Growth Obsolete?" in Moss, M. (ed.) (1973), *The Measurement of Economic and Social Performance*, National Bureau of Economic Research/Columbia University, New York.

Pearce, D., E. Barbier, and A. Markandya (1990), *Sustainable Development: Economics and Environment in the Third World*, Edward Elgar, Aldershot.

Pigou, A.C. (1932), *The Economics of Welfare*, 4th edition (1st edition 1920), Macmillan, London.

Repetto, R., W. Magrath, M. Wells, C. Beer, and F. Rossini (1989), *Wasting Assets: Natural Resources in the National Accounts*, World Resources Institute, Washington D.C.

Ruggles, R. (1983), "The United States National Income Accounts, 1947-1977: Their Conceptual Basis and Evolution" in Foss, M.F. (ed.) (1983), *The U.S. National Income and Product Accounts*, University of Chicago Press, Chicago.

Solórzano, R., R. De Camino, R. Woodward, J. Tosi, V. Watson, Vásquez, C. Villalobos, J. Jimenénez, R. Repetto, and W. Cruz (1991), *Accounts Overdue: Natural Resource Depreciation in Costa Rica*, World Resources Institute, Washington D.C.

Solow, R. (1991), "Growth Theory," in Greenaway, D., M. Bleaney, and I.M.T. Stewart (eds.) (1991), *Companion to Contemporary Economic Thought*, Routledge, London, pp. 393-415.

Tinbergen, J., and R. Hueting (1993), "GNP and Market Prices," in Goodland, R., H.E. Daly, and S. El Serafy (eds.) (1993), *Population, Technology and Lifestyle: the Transition to Sustainability*, Island Press, Washington D.C., pp. 52-62.

Tongeren, J. van, S. Schweinfest, E. Lutz, M.G. Luna, and G. Martin (1993), "Integrated Environmental and Economic Accounting: a Case Study for Mexico," in Lutz, E. (ed.) (1993), *Toward Improved Accounting for the Environment*, World Bank, Washington D.C., pp. 85-107.

United Nations Statistical Office (1992), *SNA Draft Handbook on Integrated Environmental and Economic Accounting*, Provisional Version, United Nations, New York.

World Commission on Environment and Development (1987): *Our Common Future* (The Brundtland Report), Oxford University Press, Oxford/New York..

Young, M.D. (1992), *Natural Resource Accounting: Some Australian Experiences and Observations*, Working Document 92/1 (February), CSIRO, Canberra.

CHAPTER 6

Asian Development Bank (1993), *Economic Policies for Sustainable Development*.

Commission of the European Communities (1992), *Towards Sustainability, The Fifth Environmental Action Programme*.

Confederation of Indian Industry / United Nations Development Programme / Business Council for Sustainable Development (1992), *Changing Course: Towards Sustainable Development—An Indian Industry Perspective*, New Delhi, India.

Dobson, A. (ed.) (1991), *The Green Reader*, Andre Deutsch, London, pp. 242-247.

Dunlap, R.E., G.H. Gallup, and A.M. Gallup (May 1992), *The Health of the Planet Survey: A Preliminary Report on Attitudes to the Environment and Economic Growth Measured by Surveys of Citizens in 22 Nations*, The George H. Gallup International Institute, Princeton, New Jersey, USA.

Elkington, J., and A. Dimmock (1992), *The Corporate Environmentalists: Selling Sustainable Development—But Can They Deliver?*, SustainAbility Ltd, London.

Elkington, J. and N. Robins (1994), *The UNEP Corporate Environmental Reporting Guide*, United Nations Environment Program Industry & Environment Program Activity Center (IE/OAC), Paris, France.

Fornander, K. (1991), "Japan Inc Gets into the Environment Business," *Tomorrow*, Vol. 1, No. 1, pp 38-43..

Hanneberg, P, (1991), "The Seychelles Conservation Keeps the Tourists Coming," *Tomorrow*, Vol. 1, No. 1, pp 24-37.

Hawken, P. (1993), *The Ecology of Commerce: A Declaration of Sustainability*, HarperCollins, New York.

International Institute for Sustainable Development (1992), *Sourcebook on Sustainable Development*, Manitoba, Canada.

IUCN, UNEP, and WWF (1980), *World Conservation Strategy: Living Resource Conservation for Sustainable Development,*.

IUCN, UNEP, and WWF (October 1991), *Caring for the Earth: A Strategy for Sustainable Living*, Gland, Switzerland.

MacNeill, J. (1990), "Meeting the Growth Imperative for the 21st Century," in *Sustaining Earth: Response to the Environmental Threats*, Angell, D.J.R., et al., Macmillan, London, pp. 191–205.

Meadows, D., et al. (1992), *Beyond the Limits: Confronting Global Collapse, Envisioning a Sustainable Future*, Earthscan.

Netherlands, *To Choose or to Lose: The National Environmental Policy Plan (NEPP)*, 1989; *NEPP-Plus*, 1991; *NEPP-2*, 1994..

Pearce, D., et al. (1994), *Blueprint 3: Measuring Sustainable Development*, Earthscan, London.

Pezzey, J. (1989), *Definitions of Sustainability*, CEED Discussion Paper No. 9, The UK Centre for Economic and Environmental Development, Cambridge.

Rosenberg, A.J. et al. (1993), "Achieving Sustainable Use of Renewable Resources," *Science*, Vol. 262 (5 November), pp. 828–829.

Solow, R. (1974), "The economics of resources or the resources of economics," *American Economic Review* (May): 1–14.

Solow, R. (1992), *An almost practical step toward sustainability*, Resources for the Future (40th Anniversary Lecture), Washington, D.C.

State Information Service (December 1988), *Environment and Development: The Danish Government's Action Plan*, Copenhagen, Denmark.

SustainAbility (1993a), *Coming Clean: Corporate Environmental Reporting*, , London, UK, with Deloitte Touche Tohmatsu International and the International Institute for Sustainable Development.

SustainAbility (1993b), *The LCA Sourcebook: A European Business Guide to Life-Cycle Assessment*, London, UK, with the Society for the Promotion of LCA Development (SPOLD) and Business in the Environment (BiE).

Swedish Environmental Protection Agency (1993), *Strategy for Sustainable Development*, Solna, Sweden.

Timberlake, L, "What is Sustainability," in *Green Pages: The Business of Saving the World*, Elkington J., J. Hailes, and T. Burke (eds.) (1988), Routledge, London, pp. 50–51.

United Kingdom, *This Common Inheritance: Britain's Environmental Strategy*, 1990; *First Year Report, 1991; Second Year Book, 1992*; and *The UK Strategy for Sustainable Development*, 1994.

Williams, M.J. and P.L Petesch (September 1993), *Sustaining the Earth: Role of Multilateral Development Institutions*, The Network, Center for Our Common Future, Geneva, p. 3.

World Bank (1993), *The World Bank and the Environment: Fiscal 1993*, Washington D.C.

World Commission on Environment and Development (1989), *Our Common Future*, Oxford University Press.

World Resources Institute, The Brookings Institution and the Sante Fe Institute (May 1993), *The 2050 Project: Transition to Sustainability in the Next Century*, (10-93.DSC).

Worldwatch Institute (1994), *State of the World 1994*, A Worldwatch Institute Report on Progress Towards a Sustainable Society, W.W. Norton, New York.

CHAPTER 7

Ahmad, Y., S. El Serafy, and E. Lutz (eds.) (1989), *Environmental accounting*, The World Bank, Washington D.C.

Brown, L.B. *et al.* (1994), *State of the world: 1994*, Worldwatch Institute, Washington D.C.

Cernea, M. (1993), "The Sociologist's approach to sustainable development," *Finance and Development*, 30(4): pp. 11-13.

Daily, G.C. and P.R. Ehrlich (1992), "Population, sustainability and the earth's carrying capacity," *BioScience* 42(10): pp. 761-771.

Daly, H.E., and J. Cobb (1989), *For the Common Good,* Beacon Press, Boston.

Dasgupta, P., and C. Heal (1979), *Economic Theory and Exhaustible Resources,* Cambridge University Press, Cambridge.

Ehrlich, P., and A. Ehrlich (1989a), "Too many rich folks," Populi 16(3): 3-29.

Ehrlich, P., and A. Ehrlich (1989b), "How the rich can save the poor and themselves," *Pacific and Asian Journal of Energy* 3: pp. 53-63.

Ehrlich, P.R., and J.P. Holdren (1974), "Impact of population growth," *Science* 171: pp. 1212-1217.

El Serafy, S. (1991), "The environment as capital," in Costanza, R. (ed.) *Ecological Economics*, Columbia University Press, New York, pp. 168–175.

El Serafy, S. (1993), *Country macroeconomic work and natural resources*, The World Bank, Environment Working Paper No. 58, Washington D.C.

Goodland, R., and H.E. Daly (1993a), "Why Northern income growth is not the solution to Southern poverty," *Ecological Economics* 8: pp. 85–101.

Goodland, R., and H.E. Daly (1993b), *Poverty alleviation is essential for environmental sustainability*, The World Bank, Environment Working Paper No. 42, Washington, D.C.

Hardin, G. (1993), *Living Within Limits*, Oxford University Press, New York.

Hicks, Sir J.R. (1946), *Value and Capital*, Clarendon Press, Oxford.

Ludwig, D. (1993), "Uncertainty, resource exploitation, and conservation: Lessons from history," *Science* 260 (2 April): 17, p. 36.

Lutz, E. (ed.) (1993), *Toward improved accounting for the environment*, An UNSTAT-World Bank symposium, The World Bank, Washington, D.C.

Meadows, D., D. Meadows, and J. Randers (1992), *Beyond the Limits*, Chelsea Green Publishing, Post Mills, Vermont.

Mies, M. (1991), *Consumption patterns of the North: the cause of environmental destruction and poverty in the South: women and children first*, UNCED, UNICEF and UNFPA, Geneva.

Orr, D.W. (1992), *Environmental Literacy: Education and the transition to a postmodern world*, State University of New York, Albany.

Parikh, J., and K. Parikh (1991), *Consumption patterns: the driving force of environmental stress*, UNCED, Geneva.

Putnam, R.D. (1993a), "Social capital and public affairs," *The American Prospect*, 13 (Spring): pp. 1–8.

Putnam, R.D. (1993b), *Making democracy work: civic traditions in modern Italy*, (with R. Leonardi, and R.Y. Nanetti), Princeton University Press, Princeton.

Serageldin, I. (1993a), "Making development sustainable," *Finance and Development* 30(4): pp. 6-10.

Serageldin, I. (1993b), *Development partners: Aid and cooperation in the 1990s*, SIDA, Stockholm.

Serageldin, I., and A. Steer (1994a), "Epilogue: Expanding the Capital Stock," in Serageldin, I., and A. Steer (eds.), *Making Development Sustainable: From Concepts to Action*, Environmentally Sustainable Development Occasional Paper Series no. 2. World Bank, Washington, D.C., pp. 30-32.

Serageldin, I., and A. Steer (eds.) (1994b), *Making Development Sustainable: From Concepts to action.* Environmentally Sustainable Development Occasional Paper Series no. 2., World Bank, Washington, D.C.

Simonis, U.E. (1990), *Beyond growth: elements of sustainable development*, Edition Sigma, Berlin.

Solow, R. (1974), "The economics of resources or the resources of economics," *American Economic Review* (May): pp. 1-14.

Solow, R. (1992), *An almost practical step toward sustainability*, Resources for the Future (40th Anniversary Lecture), Washington, D.C.

Steer, A., and E. Lutz (1993), "Measuring environmentally sustainable development," *Finance & Development* 30(4): pp. 20-23.

Steer, A., and V. Thomas (forthcoming), *Promoting Development that Lasts*.

Tietenberg, T. (1990), "The poverty connection to environmental policy," Challenge 33(5).

Tietenberg, T. (1992), *Environmental and Natural Resource Economics*, HarperCollins, New York.

Tinbergen, J., and R. Hueting (1991), "GNP and market prices: wrong signals for sustainable economic success that mask environmental destruction," in Goodland, R., H. Daly and S. El Serafy (eds.), *Environmentally sustainable economic development: Building on Brundtland*, The World Bank, Environment Paper 36, Washington D.C., pp. 36-42.

Vitousek, P.M., P. Ehrlich, A. Ehrlich, and P. Matson (1986), "Human appropriation of the products of photosynthesis," *BioScience* 36: pp. 368-373.

World Bank (1990): *World Development Report*, The World Bank, Washington D.C.

World Bank (1991), *World Development Report*, The World Bank, Washington D.C.

World Bank (1992a), *Environmental Assessment Sourcebook*, The World Bank, Washington D.C..

World Bank (1992b), *World Development Report 1992: Development and the Environment*, Oxford University Press, New York.

World Bank (1993a), *The World Bank and the Environment 1993*, The World Bank, Washington D.C.

World Bank (1993b), *World Development Report 1993: Investing in health*, Oxford University Press, New York.

World Bank (1995), *World Development Report: Infrastructure for Development*, The World Bank, Oxford University Press, Washington D.C..

CHAPTER 8

Ahmad, J.J., et al. (1989), *Environmental Accounting for Sustainable Development*, The International Bank for Reconstruction and Development, Washington D.C.

Clairmonte, F. (1975a), "Bananas," in Payer, C. (ed.), *Commodity Trade of the Third World*, Macmillan, London.

Clairmonte, F. (1975b), "Dynamics of International Exploitation," *Economic and Political Weekly*, vol. X.

Henderson, H. (1991), *Paradigms in Progress* , Knowledge Systems, Indiana, Indianapolis.

ICVA and Eurostep (1993), *The Reality of Aid*, United Kingdom.

Khor, M. (1993), "Towards a new North-South economic dialogue," *Third World Resurgence*, no. 36 (August).

Khor Kok Peng (1983), *The Malaysian Economy: Structures and Dependence*, Institut Masyarakat, Penang.

Khor Kok Peng (1994), *The End of the Uruguay Round and Third World Interests*, Third World Network, Penang.

Panayotou, T. (1993), *Green Markets*, Institute for Contemporary Studies, San Francisco, California.

Papic, A. (1991), "Net Transfer of Financial Resources from the South," *South Letter*, no. 11 (September).

Pearce, D.W., and J.J. Warford (1993), *World Without End*, Oxford University Press, Oxford.

The International Bank for Reconstruction and Development (1990), *World Development Report 1990, Poverty*, Oxford University Press, Oxford.

The International Bank for Reconstruction and Development (1993a), *The East Asia Miracle*, Oxford University Press.

The International Bank for Reconstruction and Development (1993b), *World Development Report 1993, Investing in Health*, Oxford University Press.

UNCED Secretariat (1991), "The International Economy and Environment and Development," Paper A/CONF 151/PC/47.

UNCTAD (1981), *Trade and Development Report 1981*, United Nations, New York.

United Nations (1990), *World Economic Survey 1990*, New York.

United Nations (1991), *World Economic Survey 1991*, New York.

United Nations (1993), *World Economic Survey 1993*, New York.

United Nations (1993a), *Agenda 21: The UN Program of Action from Rio*, New York.

CHAPTER 9

Anderson, V., *Alternative Economic Indicators*, New Economics Foundation, Routledge, London and New York (1991).

Bosch, P., et al., *Methodological Problems in the Calculation of Adjusted National Income Figures*, CBS the Netherlands, 1992–1996, European Union, DGXII, Research and development in the field of the environment, phase II, area III, socioeconomic environmental research, research task III, 1, 2, 2.

Carey, M. (1981), *Social Measurement and Social Indicators, Issues of Policy and Theory*.

Cobb C., and J. Cobb, Jr., eds. (1994),The Green National Product: A proposed Index of Sustainable Economic Welfare, University Press of America, New York (1992).

Cobb, C., and T. Halstead (1994), *The Genuine Progress Indicator: A Measure of the Socio-Economic Well-Being and Sustainability of the Nation: 1950-1992*, Redefining Progress, California.

Daly, H., and J. Cobb (1990), *For the Common Good*, Beacon Press, Boston.

Diefenbacher, H. (1994), "The Index of Sustainable Economic Welfare: A Case Study of the Federal Republic of Germany," in Cobb and Cobb, 1994.

Diefenbacher, H., and S. Habicht-Erenler (1991), *Wohlstand und Wachstum—Neuere Konzepte zur Erfassung von Sozial—und Umweltverträglichkeit*.

Ekins, P. (1992), *Wealth Beyond Measure: An Atlas of New Economics*, London.

Hareide, D. (1990), *The Good Norway* (Det Gode Norway), Gyldendal.

Henderson, H. (1991), *Paradigms in Progress—Life beyond Economics*, Knowledge Systems, Indianapolis, Indiana.

Jackson, T., and N. Marks (1994), *Measuring Sustainable Economic Welfare—A Pilot Index: 1950-1990*, SEI (Stockholm Environment Institute), published in cooperation with The New Economic Foundation/UK, Stockholm.

Jesperson, J. (1994), "A Welfare Index for the Danish Economy," presented at the ECS-meeting in Wuppertal, 15 June (draft form).

Kelley, A.C., "The Human development Index: Handle with Care," in: *Population and Development Review* 17(2), p. 315-24.

Klingebiehl, S. (1992), *Entwicklungsindikatoren in der politischen und wissenschaftlichen Diskussion*, INEF Report, Heft 2.

Meyer, C.A., *Environmental and natural resource accounting: Where to begin?*, World Resources Institute.

Nohlen D./F. Nuscheler, "Indikatoren von Unterentwicklung und Entwicklung," in: Nohlen/Nuscheler (eds.) (1993), *Handbuch der Dritten Welt*, Bonn.

Obermayr, B, K. Steiner, E. Stockhammer, and H. Hockrieber (1994), *Die Entwicklung des ISEW in Osterreich von 1955 bis 1992* (forthcoming).

OECD (1994), *Natural resource accounts: taking stock in the OECD countries*, Paris.

Rosenberg, D., and T. Oegema (1994), *A Pilot Index of Sustainable Economic Welfare for the Netherlands, 1950-1992*, Institute for Environment and Systems Analysis, Amsterdam (forthcoming).

Seifert, E.K. (1989), "Aristotelian Justice in modern economics," in Greek Philosophical Society (ed.), *On Justice: Plato's and Aristotle's conception in relation to modern and contemporary theories of justice*, Athens.

United Nations Development Program (1994), *Human Development Report (HDR)*, p. 91.

CHAPTER 10

Daly, H., and J. Cobb (1990), *For the Common Good*, Beacon Press, Boston.

Estes, R.J., (1988), *Trends in World Social Development: the Social Progress of Nations, 1970-1987*, Praeger, New York.

Henderson, H. (1981), *Politics of the Solar Age*, Doubleday, New York.

Henderson, H. (1991), *Paradigms in Progress* , Knowledge Systems, Indiana, Indianapolis.

Redefining Wealth & Progress: New Ways to Measure Economic, Social and Environmental Change, Knowledge Systems, Indianapolis, Indiana, 1990.

Sivard, R.L. (1991), *World Military and Social Expenditures*, World Priorities Institute, 14th Edition.

South Commission (1990), *Challenge to the South*, Oxford University Press, New York.

INTRODUCTION PART IV

Ahmad, Y.J., S. El Serafy, and E. Lutz (eds.) (1989), *Environmental Accounting for Sustainable Development*, A UNEP-World Bank Symposium, The World Bank, Washington D.C.

Lutz, E. (1993), "Toward Improved Accounting for the Environment: an Overview," in Lutz, E. (ed.) (1993), *Toward Improved Accounting for the Environment*, World Bank, Washington D.C., pp. 22-45.

Pearce, D.W., and R.K. Turner (1990), *Economics of Natural Resources and the environment*, Harvester Wheatsheaf, New York.

Repetto, R., W. Magrath, M. Wells, C. Beer, and F. Rossini (1989), *Wasting Assets: Natural Resources in the National Accounts*, World Resources Institute, Washington D.C.

CHAPTER 11

Daly, H.E. (1989), "Toward a Measure of Sustainable Social Net Product," in Ahmad, Y.J. *et al.* (eds.), *Environmental and Natural Resource Accounting and Their Relevance to the Measurement of Sustainable Development*, The World Bank, Washington D.C.

Daly, H.E., and J.B. Cobb (1990), *For the Common Good*, Green Print, Merlin Press, London.

Hueting, R. and C. Leipert (1990), "Economic Growth, National Income and the Blocked Choices for the Environment," in *The Environmentalist*, Vol. 10, No. 1, pp. 25-38.

Leipert, C. (1989), *Die heimlichen Kosten des Fortschritts. Wie Umweltzerstörung das Wirtschaftswachstum fördert (The Hidden Costs of Progress: How Environmental Desctruction Furthers Increasing Economic Growth)*, S. Fischer Verlag, Frankfurt/M.

Leipert, C. (1989), "Social Costs of the Economic Process and National Acccounts. The Example of Defensive Expenditures," in *Journal of Interdisiplinary Economics*, Vol. 3, No. 1, pp. 27-46.

Leipert, C. (1989), "National Income and Economic Growth: The Conceptual Side of Defensive Expenditures," in *Journal of Economic Issues*, Vol. 23, No. 3, pp. 843-856.

Olson, M. (1977), "The Treatment of Externalities in National Income Statistics," in Wingo, L., and A. Evans (eds.), *Public Economics and the Quality of Life*, John Hopkins University Press, Baltimore, Maryland.

Chapter 12

Ahmad, Y.J., S. El Serafy, and E. Lutz (eds.) (1989), *Environmental Accounting for Sustainable Development*, A UNEP-World Bank Symposium, The World Bank, Washington D.C.

Bartelmus, P., E. Lutz, and J. van Tongeren (1993), "Environmental Accounting with an Operational Perspective," ESD and the World Bank, Washington D.C.

El Serafy, S. (1981), "Absorptive Capacity, the Demand for Revenue and the Supply of Petroleum," *Journal of Energy & Development*, Vol. VII, No. 1, (Autumn).

El Serafy, S. (1989), "The Proper Calculation of Income from Depletable Natural Resources," in Ahmad, Y.J., S. El Serafy, and E. Lutz (eds.), *Environmental Accounting for Sustainable Development*, The World Bank, Washington, D.C.

El Serafy, S. (1991), "The Environment as Capital," in Costanza, R. (ed.), *Ecological Economics*, Columbia University Press, New York.

El Serafy, S. (1993), "Depletable Rersources: Fixed Capital or Inventories?" in Franz, A., and C. Stahmer (eds.), *Approaches to Environmental Accounting*, Physica-Verlag, Heidelberg.

Hartwick, J.M., and A. Hageman (1993), "Economic Depreciation of Mineral Stocks and the Contribution of El Serafy," in Lutz, E. (ed.), *Toward Improved Accounting for the Environment*, UNSTAT-World Bank Symposium, The World Bank, Washington D.C.

Hicks, J.R. (1946), *Value and Capital*, Chapter XIV (Income), Second Edition, Clarendon Press, Oxford.

Hicks, J.R. (1974), "Capital Controversies: Ancient and Modern," *American Economic Review*, (May).

Hueting, R. (1989), "Correcting National Income for Environmental Losses: Toward a Practical Solution," in Ahmad, Y.J., S. El Serafy, and E. Lutz (1989), *Environmental Accounting for Sustainable Development*, A UNEP-World Bank Symnposium, the World Bank, Washington D.C., p. 37.

Hueting, R. (1993), "Calculating a Sustainable National Income: A Practical Solution for a Theoretical Dilemma," in Franz, A., and C.

Stahmer (eds.) (1993), *Approaches to Environmental Accounting*, Physica-Verlag, Heidelberg.

Marshall, A. (1947), *Principles of Economics*, 8th Edition, 1920, Macmillan and Co., London. p. 167.

Repetto, R., et al. (1989), *Wasting Assets: Natural Resources in the National Income Accounts*, World Resources Institute, Washington D.C.

Solorzano, R. et al. (1991), *Accounts Overdue: Natural Resource Depreciation in Costa Rica*, World Resources Institute, Washington D.C.

Solow, R. (1992), "An Almost Practical Step Toward Sustainability," *Resources for the Future*, Washington D.C. (October).

United Nations (1993), *Integrated Environmental and Economic Accounting*, Handbook of National Accounting, Studies in Methods, Series F, No. 61, Sales No. E. 93. XVII.12, New York.

United States Department of Commerce (1994), "Accounting for Mineral Resources: Issues and BEA's Initital Estimates," *Survey of Current Business*, Bureau of Economic Analysis, April

CHAPTER 13

Daly, H.E. and Cobb, Jr., J.B. (1989), *For the Common Good*, Beacon Press, Boston.

Hennipman, P. (1962), "Doeleinden en criteria," in: *Theorie van de Economische Politiek*, Leiden, P. 150.

Hicks, J.R. (1948), "On the Valuation of Social Income," *Economica* (August).

Hueting, r. (1970), *Wat is de natuur ons waard? (What is nature worth to us?)*, Wereldvenster, Baarn.

Hueting, R. (1980a), *New Scarcity and Economic Growth*, North-Holland Publishing Company, Amsterdam, New York, Oxford (Dutch edition 1974).

Hueting, R. (1980b), "Environment and Growth: Expectations and Scenarios," in: *Prospects of Economic Growth*, Kuipers, S.K., and Lanjouw, G.J. (eds.), North-Holland Publishing Company, Amsterdam, New York, Oxford, p. 89.

Hueting, R. (1981), "The relation between growth of production and energy consumption. Does growth fanaticism blind one?," mimeo (in Dutch) in: *Economisch-Statistische Berichten* (24 June), pp. 609–611.

Hueting, R. (1986), *A Note on the Construction of an Environmental Indicator in Monetary Terms as a Supplement to National Income with the Aid of Basic Environmental Statistics*, Report to Prof. E. Salim, Minister of Population and Environment of Indonesia, Jakarta.

Hueting, R. (1987), "An Economic Scenario That Gives Top Priority to Saving the Environment," *Ecological Modelling*, Volume 38, Nos. 1/2 (September), p. 130.

Hueting, R. (1989), "Correcting National Income for Environmental Losses: Towards a Practical Solution," in Ahmad, Y., S. El Serafy, and E. Lutz (eds.), *Environmental Accounting for Sustainable Development*, The World Bank, Washington, D.C.

Hueting, R. (1991), "The Use of the Discount Rate in a Cost–Benefit Analysis for Different Uses of a Humid Tropical Forest Area," *Ecological Economics 3*.

Hueting, R. (1992), "The Economic Functions of the Environment," in Ekins, P., and M. Max-Neef (eds), *Real-Life Economics*, Routledge, London and New York.

Hueting, R., Bosch, P. and De Boer, B. (1992), *Methodology for the Calculation of Sustainable National Income*, Netherlands Central Bureau of Statistics, Statistical Essays, M44, SDU/Publishers, Gravenhage; also published as WWF report.

Hueting, R. (1993), "Calculating a Sustainable National Income: A Practical Solution for a Theoretical Dilemma," in Franz, A. and Stahmer, C. (eds.), *Approaches to Environmental Accounting: Proceedings of the IARIW Conference on Environmental Accounting, Baden (near Vienna), Austria, 27–29 May 1991*, Physica-Verlag, Heidelberg.

Hueting, R. (1994), "Three persistent myths in the environmental debate," *Economisch-Statistische Berichten* (25 November).

Kuznets, S. (1948), "On the Valuation of Social Income," *Economica* (February/May).

Marcuse, R. (1964), *One-Dimensional Man: The Ideology of Industrial Society*, London.

Pigou, A.C. (1949), *Income,* London.

Potma, T. (1990), "A strategy for change," paper presented at the 10th Annual International Conference on "Strategic Bridging," Stockholm, September 24–27.

Robbins, L. (1952), *An Essay on the Nature and Significance of Economic Science,* Second edition, London, pp. 24, 38, and 145.

Tinbergen, J., and R. Hueting (1991), "GNP and Market Prices: Wrong Signals for Sustainable Economic Success That Masks Environmental Destruction," in Goodland, R., H. Daly, S. El Serafy, and B. von Droste, *Environmentally Sustainable Economic Development: Building on Brundtland,* United Nations Educational, Scientific, and Cultural Organization, Paris.

Also published in Goodland, R., *et al.* (eds.) (1991), *Environmentally Sustainable Economic Development: Building on Brundtland,* Environment Working Paper No. 46, The World Bank, Washington, D.C.

Also published in Goodland, R. *et al.* (eds.) (1992), *Population, Technology and Lifestyle: The Transition to Sustainability,* Island Press, The International Bank for Reconstruction and Development and UNESCO, Washington D.C.

CHAPTER 14

Ahmad, Y., S. El Serafy, and E. Lutz (eds.) (1989), *Environmental Accounting for Sustainable Development,* A UNEP-World Bank Symposium, World Bank, Washington, D.C.

Bartelmus, P., C. Stahmer, and J. van Tongeren (1991), "Integrated environmental and economic accounting: framework for a SNA satellite system," in: *Review of Income and Wealth,* Series 37, No. 2, pp. 111–148.

Costanza, R. (ed.) (1991), *Ecological Economics—The Science and Management of Sustainability, Part II: Accounting, Modeling and Analysis,* Irvington, New York.

Daly, H. (1991), "Elements of environmental macroeconomics," in: Costanza, R. (ed.), *Ecological Economics. The Science and Management of Sustainability,* New York, pp. 32–46.

El Serafy, S. (1993), "Depletable Resources: Fixed Capital or Inventories," in: Franz, A. and C. Stahmer (eds.), *Approaches to Environmental Accounting,* Physica-Verlag, Heidelberg, pp. 245–258.

Franz, A., and C. Stahmer (1993), *Approaches to Environmental Accounting*, Proceedings of the IARIW Conference on Environmental Accounting, Baden, Austria, May 1991, Physica-Verlag, Heidelberg.

Harrison, A., and P. Hill (1994), *Accounting for Subsoil Assets in the 1993 SNA and a Corrigendum*, paper presented at the Meeting on National Accounts and the Environment, London, March 1994.

Hueting, R. (1980), *New Scarcity and Economic Growth: More Welfare through less Production?* Amsterdam, New York, Oxford.

Hueting, R. (1993) "Calculating a sustainable national income: A practical solution for a theoretical dilemma," in: Stahmer, F.(1993), pp. 39—69.

Leipert, C. (1991), *The Role of Defensive Expenditures in a System of Integrated Economic-Ecological Accounting*, Internationales Institut für Umwelt und Gesellschaft, Berlin.

Lutz, E. (ed.) (1993), *Toward Improved Accounting for the Environment*, A UNSTAT-World Bank Symposium, World Bank, Washington, D.C.

NNW Measurement Committee (1973), *Measuring Net National Welfare of Japan*, Tokyo.

Nordhaus, W.D., and J. Tobin (1973), "Is growth obsolete?" in: Moss, M. (ed.), "The Measurement of Economic and Social Performance," *Studies in Income and Wealth*, vol. 38, New York/London, pp. 509-531.

OECD (1989), *Environmental Policy Benefits—Monetary Valuation*, Study prepared by D.W. Pearce and A. Markandya, Paris.

Repetto, R., et al. (1989), *Wasting assets. Natural Resources in the National Income Accounts*, World Resources Institute, Washington, D.C.

United Nations (1991), *Official Records of the Economic and Social Council*, 1991, Supplement No. 5 (E/1991/25), para. 154 (3)(iv).

United Nations (1992), *Report of the United Nations Conference on Environment and Development, Rio de Janeiro, 3-14 June 1992, Vol. I. Resolutions adopted by the Conference*, Sales No. E.93, I.8 New York.

United Nations (1993), *Integrated Environmental and Economic Accounting*, Handbook of National Accounting, Studies in Methods, Series F, No. 61, Sales No. E. 93. XVII.12, New York.

United Nations (1994), *System of National Accounts*, Studies in Methods, Series F, No. 2, Rev. 4, Sales No. E 94. XVII.4, New York.

PART V

Ekins, P., and M. Max-Neef (1992), *Real life economics, understanding wealth creation*, Routledge, London.

Giarini, O. and W.R. Stahel (1993), *The Limits to Certainty: Facing Risks in the New Service Economy*, Progress, Geneva.

Grubb, M., J. Edmonds, P. ten Brink and M. Morrison (1993), "The costs of Limiting Fossil-Fuel CO2 Emissions: A Survey and Analysis," in: *Annu. Rev. Energy Environ.*, 18: pp. 397–478.

Prognos (1992), *Externkosten der Energieerzeugung. Gutachten für den Bundesminister für Wirtschaft*, Bonn.

Repetto, R., R.C. Dower, R. Jenkins, and J. Geoghean (1992), *Green Fees: How a Tax Shift Can Work for the Environment and the Economy*, World Resources Institute, Washington D.C.

Schmidt-Bleek, F. (1994), *Wieviel Umwelt braucht der Mensch?* Birkhäuser, Basel.

Serageldin, I. (1993a), "Making Development Sustainable," *Finance and Development* 30(4): pp. 6–10.

Sheng, F. (1995), *Making ends meet, An overview of global efforts to assess true economic progress*, WWF-International.

Weizsäcker, E.U. von, and J. Jesinghaus (1992), *Ecological Tax Reform. Policy Proposal for Sustainable Development*, Zed Books, London.

Weizsäcker, E.U. von (1994), *Earth Politics*, Zed Books, London (also available in Japanese, German and Spanish).

Index